Arduino Programming

Step-by-Step Guide To Master Arduino Hardware and Software

Mark Torvalds

Copyright 2018 Mark Torvalds - All rights reserved.

If you would like to share this book with another person, please purchase an additional copy for each recipient. Thank you for respecting the hard work of this author. Otherwise, the transmission, duplication or reproduction of any of the following work including specific information will be considered an illegal act irrespective of if it is done electronically or in print. This extends to creating a secondary or tertiary copy of the work or a recorded copy and is only allowed with express written consent from the Publisher. All additional right reserved.

TABLE OF CONTENTS

Chapter 1	10
Introduction to Arduino	10
Introduction to Embedded Systems	11
	11
Examples of embedded systems	11
Consumer electronics	13
Networking	14
What is an Arduino?	14
Why use an Arduino?	15
Arduino Boards	17
General Overview of Arduino UNO	18
Memory Types on ATMEGA 328P (the heart of Arduino UNO)	21
Questions	23
Chapter 2	24
Hardware & Tools	24
Questions	39
Chapter 3	41
Getting started with Arduino	41
First example: LED blinking (Wiring)	42
Steps	42
First example: LED blinking (Coding)	43
Steps	43
Example 2 (Wiring)	50
Example 2 (Coding)	51

Example 3 (Wiring) LED blinking using two push buttons 56

Example 3 (Coding) 57

Questions 62

Chapter 4 63

Inputs, Outputs, and Sensors 63

• Digital signal: 64

• Analog signal: 64

Why analog signals are important? 65

How sensors generate the analog signals? 67

Steps 68

Example 4: Control light amount using potentiometer (wiring) 72

Example 4: Control light amount using potentiometer (Coding) 74

Example 5: Photo resistor as light sensor (Components) 77

Example 5: Photo resistor as light sensor (Wiring) 78

Example 5: Photo resistor as light sensor (Coding) 78

What is the pulse width modulation? 81

How we can use it? 82

Example 6: LED with PWM (wiring) 83

Example 6: LED with PWM (coding) 83

Questions 86

Chapter 5 87

Computer interfacing with Arduino 87

FTDI Chips 88

Example7: Temperature sensor with serial interface (Components) 89

Example7: Temperature sensor with serial interface (Wiring) 90

Example7: Temperature sensor with serial interface (Coding) 90

Example 8: Showing the strength of the LED Light on the serial monitor (Wiring) 96

Example 8: Showing the strength of the LED Light on the serial monitor (Coding) 96

Example 9: Turn on / off your LED using your computer (Components) 99

Example 9: Turn on / off your LED using your computer (Coding) ... 101

Questions 104

Chapter 6 106

The Motors 106

Intro 107

Example 10: Using the direct current motor "DC Motor" (Components) 108

Example 10: Using the direct current motor "DC Motor" (Wiring) 109

Example 11: Using the direct current motor "Servo Motor" (Components) ....

Example 11: Using the direct current motor "Servo Motor" (Wiring) .. 112

Example 11: Using the direct current motor "Servo Motor" (Coding) . 112

Questions 115

Chapter 7 116

Advanced Inputs and Outputs 116

Intro 117

Example:10 16x2 LCD interfacing(Components) 121

Steps 121

Use the potentiometer to control the brightness of the display 128

Interface the Keypad with the Arduino 143

Example 11: Using the keypad with Arduino (Components) 146

Example 11: Using the keypad with Arduino (Wiring) 147

Example 11: Using the keypad with Arduino (Coding) 149

Introduction to relays	152
What is a relay?	153
Questions	157
Chapter 8	**158**
Arduino shields	**158**
Intro to shields	158
Questions	168
Chapter 9	**169**
Final Project	**169**

Chapter 1

Introduction to Arduino

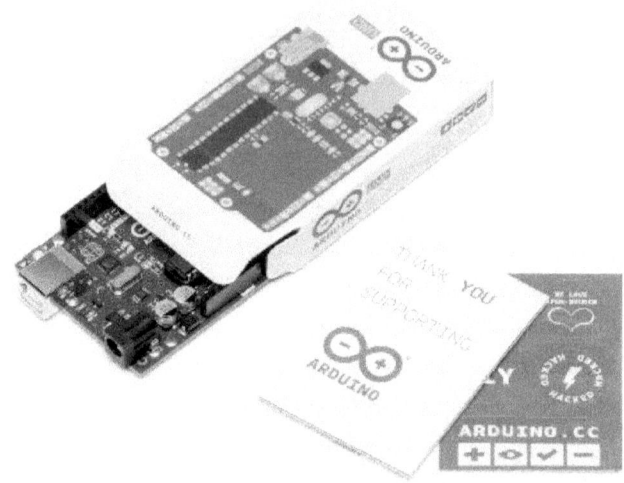

What you will learn in this chapter:

📇Introduction to Embedded Systems

📇What is an Arduino?

📇Why use an Arduino?

📇Arduino boards

📇General overview of Arduino UNO

What you will need for this chapter:

An Arduino UNO board

Introduction to Embedded Systems

An embedded system is a computer system (including hardware and software) that is designed for a special purpose and usually has no graphical user interface. It can be a microprocessor or a microcontroller, which is the heart of the system. The difference between a microcontroller and microprocessor is that the microcontroller contains a microprocessor and also peripherals such as flash, RAM, etc., but on the other hand the microprocessor only implements the central processing unit (CPU).

Figure 1.1: Description of the microcontroller

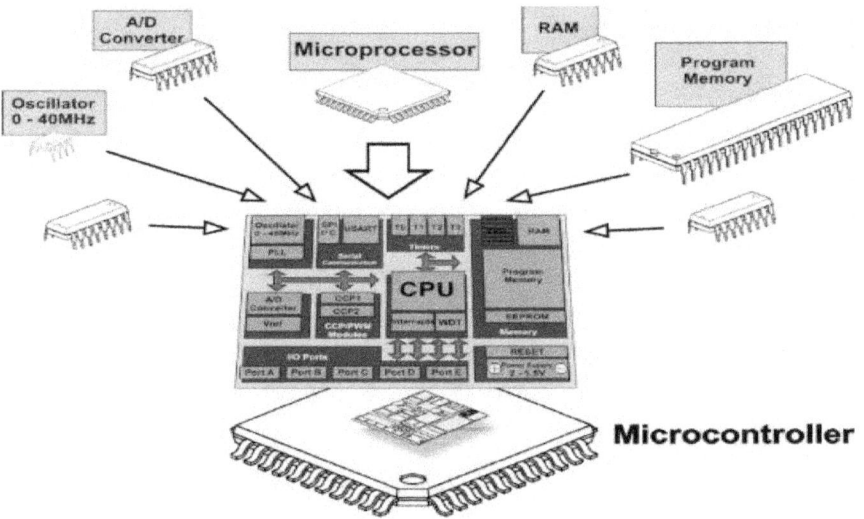

Examples of embedded systems:
- Embedded systems are widely used in many devices and applications like:

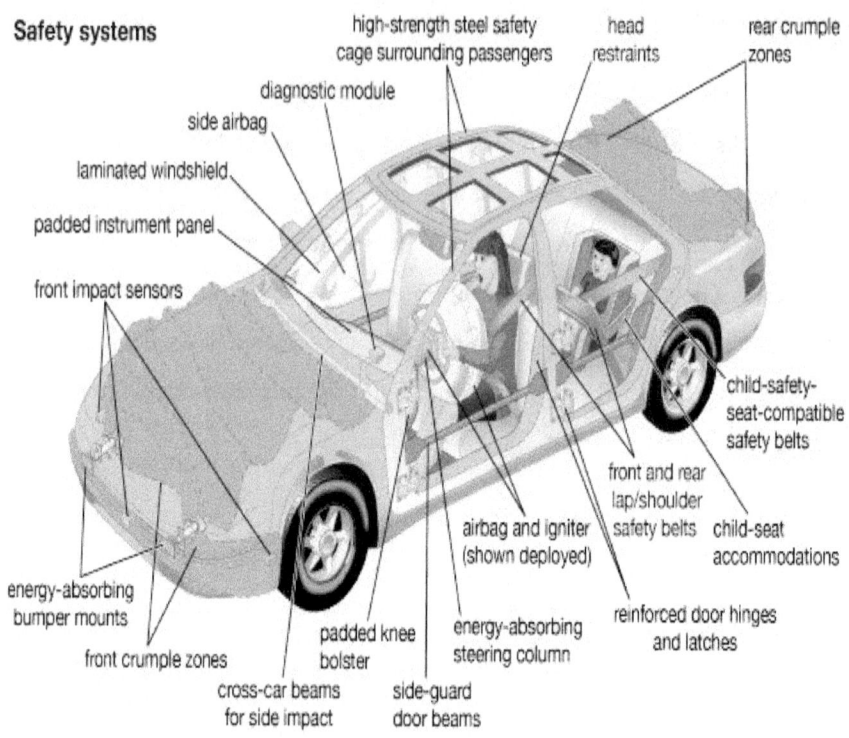

Figure 1.2: Examples of automotive systems

Consumer Electronics

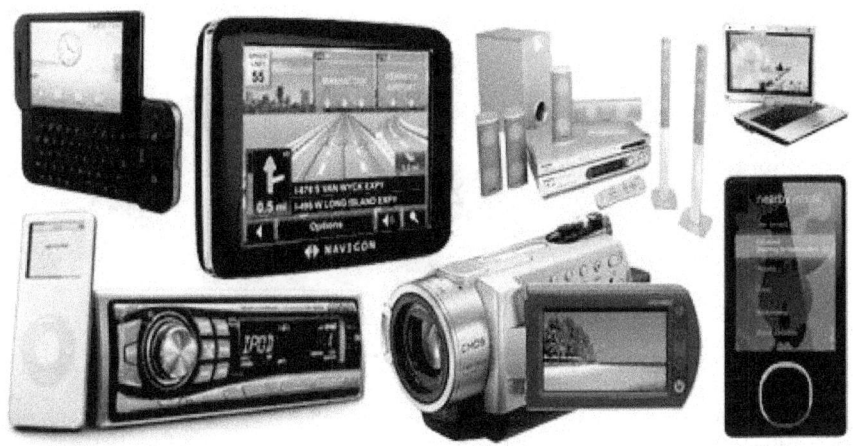

Figure 1.3: Examples of consumer electronics

- Digital and Analog Televisions

- Set Top Boxes (DVDS, VCRs, Cable Boxes, etc.)

- Personal Data Assistants (PDAs)

- Cameras

- Global Positioning System (GPS)

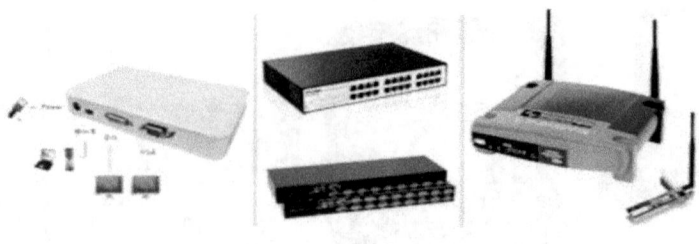

Figure 1.4: Examples of networking devices

- Routers

- Switches

- Gateways & Hubs

What is an Arduino?

After viewing the concepts of embedded systems and microcontrollers, it's now time to know exactly what an Arduino is.

An Arduino is an open source microcontroller platform, and open source here means that you can get the internal designs and the schematics of the Arduino and also make derivatives of Arduino boards or entirely new products based on Arduino technology.

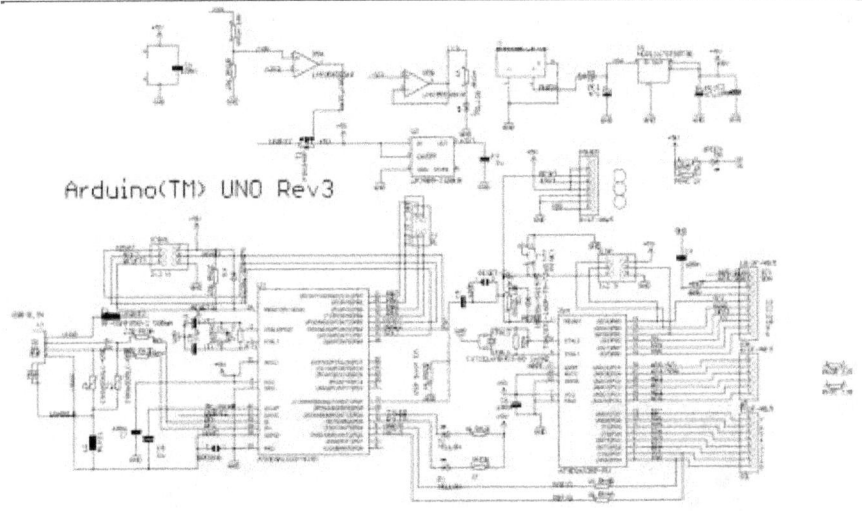

Figure 1.5 schematic of Arduino UNO Rev3

Why use an Arduino?

- **Arduinos** have many advantages that make it very popular for hobbyists and engineers.

- **Easy to use**: You can program it using a USB cable and you don't need to care about the burner like with other microcontrollers.

- **Large and supporting community**: There is a huge, supportive community that can help you to develop any project.

- **Arduino Libraries**: You can use a lot of libraries in the Arduino environment to save time when writing code.

- **Arduino Language:** It's so easy to learn the Arduino programming language, which developed by the Italian team in 2005. The language is derived from C.

Arduino Boards

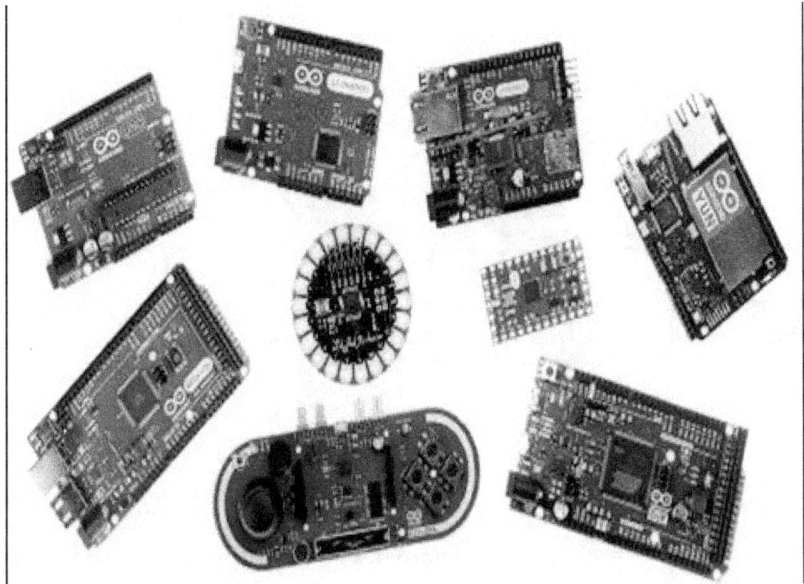

Figure 1.6: Arduino Boards

The differences between Arduino boards are:

- The number of input and output pins.

- The type of microprocessor on the board.

- The number of built-in parts on the board.

- **But we will use and examine the most popular board, which is the Arduino UNO.**

General overview of Arduino UNO

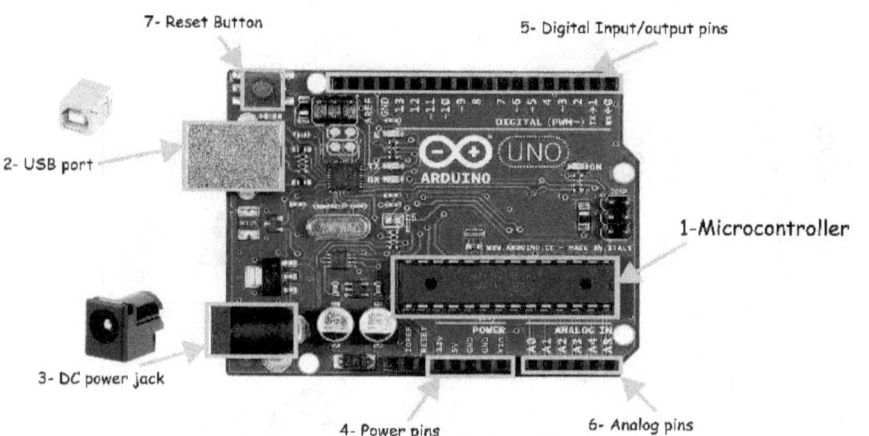

1. **Microcontroller:** This is the heart of the Arduino. Arduino UNO and most other boards contain an Atmel microcontroller unit (MCU) and use an AVR microcontroller. The Arduino UNO here uses an ATMega 328p. This microcontroller is responsible for processing all compiled code and the execution of all commands. The Arduino programming language makes it so easy to access all the peripherals like the analog to digital converter (ADCs), general purpose input/output pins (GPIO), and also it contains 16MHz crystal oscillator.

2. **USB Port:** Used to connect the Arduino to the computer and provide the 5v power to turn on the board.

3. **DC power jack:** When you're ready to unplug your project from the computer, you have other power options like the DC power jack.

4. **Power pins:** The Arduino has two main regulators:

- 5v for digital input/output.

- 3.3v for used when you connect shields and external circuitry.

   And also two pins for the ground.

5. **Digital Input/Output pins**: The most important part that we will care about during your projects is the general–purpose input/output pins. We will use it via the programs. They can serve as an input or output, and they also have other special function like pulse width modulation (PWM).

6. **Analog Pins**: The ADC pins act as analog inputs to measure voltage between 0v and 5v.

7. **Reset Button**: This is used to reload the program on the Arduino board.

Memory Types on ATMEGA 328P (the heart of Arduino UNO)

- **SRAM**: The memory that used to temporarily store the variables.

- **Flash Disk**: It is a storage area that is used to store the program that makes the microcontroller work.

- **EEPROM**: Its responsibility is to permanently store some variables, similar to your hard disk in the PC.

- **Bootloader**: Its functionality is to enable programming via USB with no external hardware.

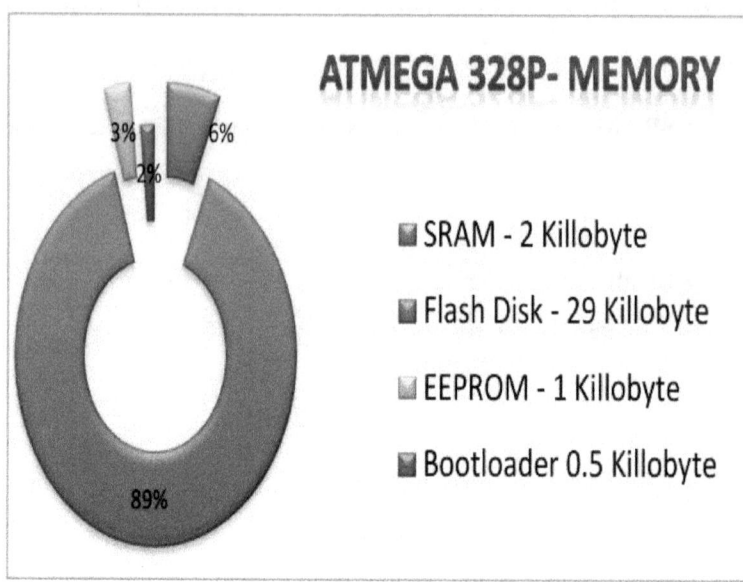

Questions

1. What is an embedded system?

2. What is the difference between a microprocessor and a microcontroller?

3. What is the purpose of GPIO pins on the Arduino board?

4. How many memory types, and their sizes, are on the Atmega328p?

Chapter 2

Hardware & Tools

What you will learn in this chapter:

▣ Prepare and understand your hardware

▣ Install the Arduino IDE

What you will need for this chapter:

▣ An Arduino UNO board   ▣ USB cable

▣ Breadboard   ▣ LEDs

▣ Resistors   ▣ Multimeter

▣ Wires

An Arduino UNO rev.3

● This is the board that contains the ATMega328P microcontroller. It has 14digital input/output pins including 6 pins can be used as PWM outputs. It has a 16MHz resonator, a USB connection, a power jack, and more. (for more details you can review Chapter 1).

Rev3 features:

●The reset circuit is stronger than the older revisions.

● It includes ATMega 16U2 instead of 8u2.

Figure 2.1: Arduino UNO Rev3

A-B USB Cable

• This cable is used to connect your Arduino to your computer. You can buy any type but I prefer them to be as short as possible.

Figure 2.2: A-B USB Cable

Breadboard

• We will use this board to connect the components together without soldering, you can use any type.

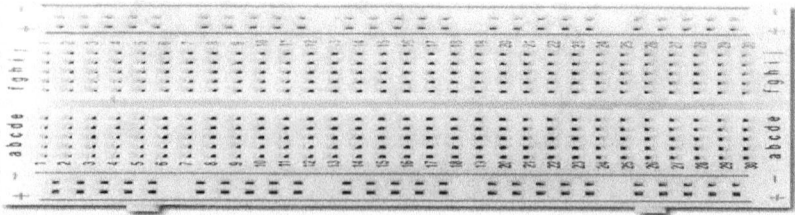

LEDs

- LED stands for light emitting diode. LED are like bulbs and they are available in many types and colors like red, green, yellow, white, or orange. They are mainly used for debugging purposes.

We will need:

- At least 10 LEDs.

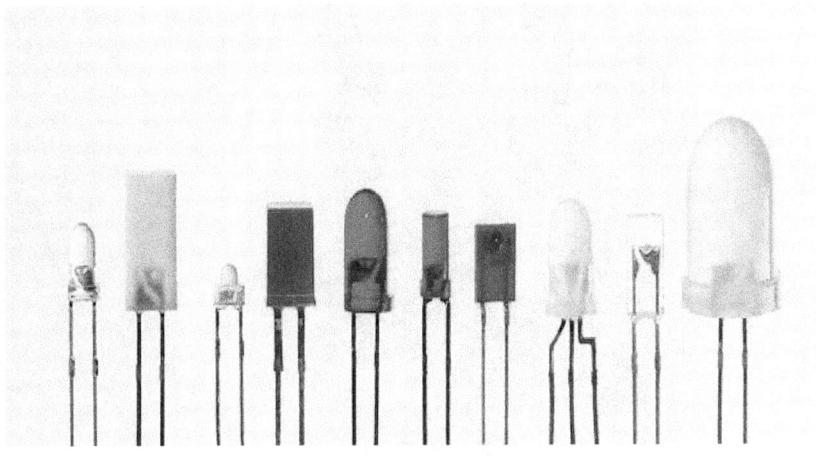

Figure 2.4: Different types of LEDs

Multimeter device (optional)

• It's an electronic device that is used to measure voltage, current, resistant, capacitance.

We will need:

- Autoranging multimeter (1)

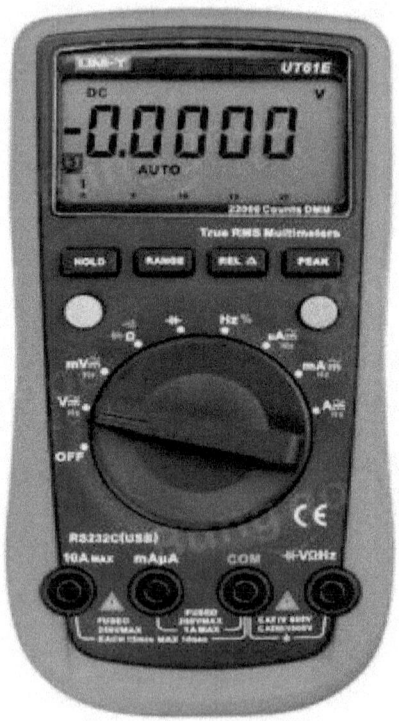

• Autoranging means that multimeter can detect the measurements range automatically.

Resistors

• Resistors are an electrical component that is used to control the flow of current in a circuit.

We will need:

- 560 ohm resistors (5)

- 10 k ohm resistors (5).

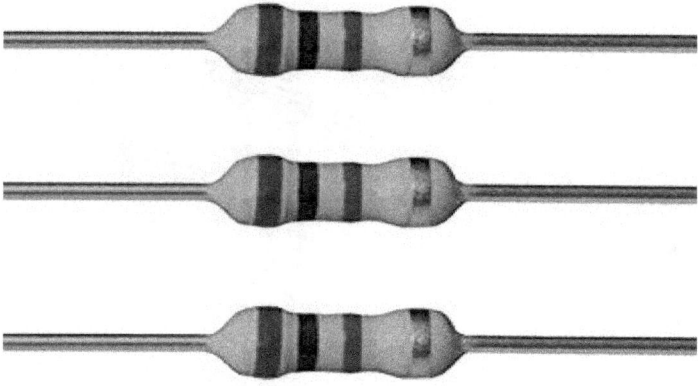

Figure 2.6: Resistor

Wires

- Jumper wires are used to connect our components with each other on the breadboard.

We will need:

Male to Male

Female to Male

Female to Female

The Arduino IDE

- The Arduino integrated development environment is the tool that will be used to write and upload code on our Arduino.

- It uses a very simple programming language which is Arduino C.

Install Arduino IDE on Linux

- You can install the Arduino IDE from the software center in Ubuntu.

- Write "Arduino" on the search form, then click enter/install.

- If you use any other Linux distro, you can search for the Arduino IDE on its software center.

Install Arduino IDE on Mac/Windows

- Go to https://www.arduino.cc/en/main/software .

- Select the windows/Mac installer.

After the installation

- Connect the cable to the Arduino board.

- Now open the Arduino IDE.

AN OPEN PROJECT WRITTEN, DEBUGGED,
AND SUPPORTED BY ARDUINO.CC AND
THE ARDUINO COMMUNITY WORLDWIDE

LEARN MORE ABOUT THE CONTRIBUTORS
OF **ARDUINO.CC** on arduino.cc/credits

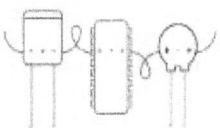

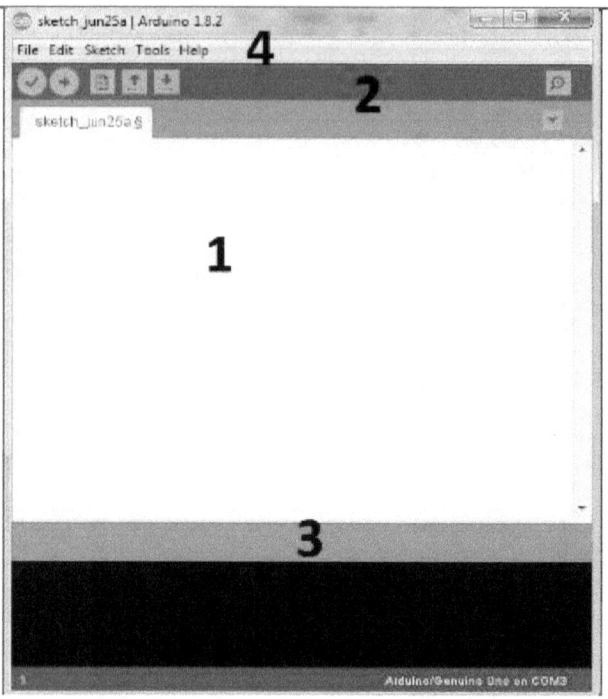

1.
   he code will be written in this area.

2.
   avigation bar to upload , verify, save

3.
   onsole area to show errors and warning.

4.
   enu bar.

Prepare the Arduino IDE

- irst we will go to the tools menu, then choose Arduino UNO.

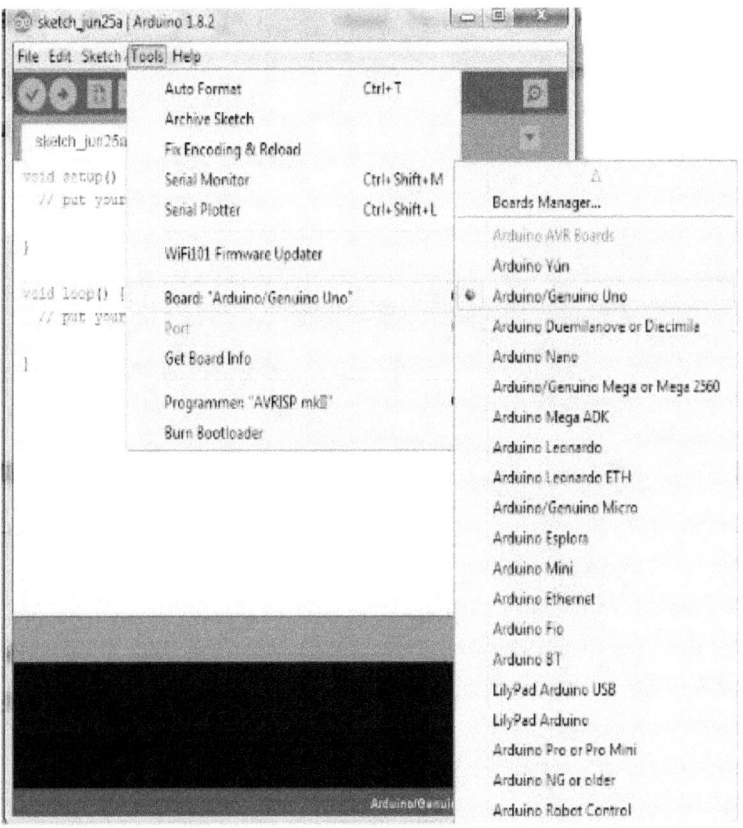

- Second, we will select Tools, Port, then Com ("x").

X is the port number.

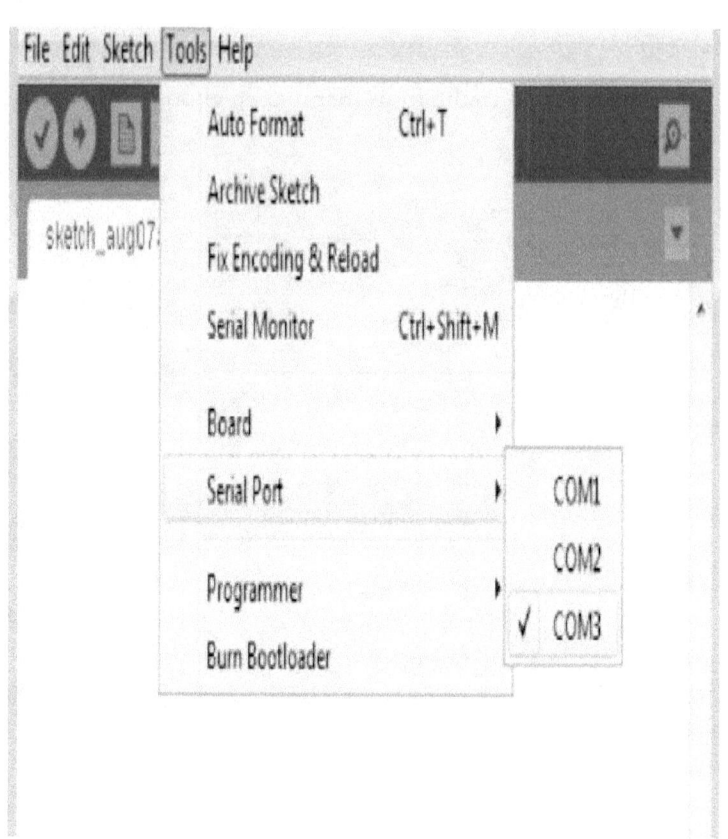

- After that, you can start writing your first program by selecting file > new.

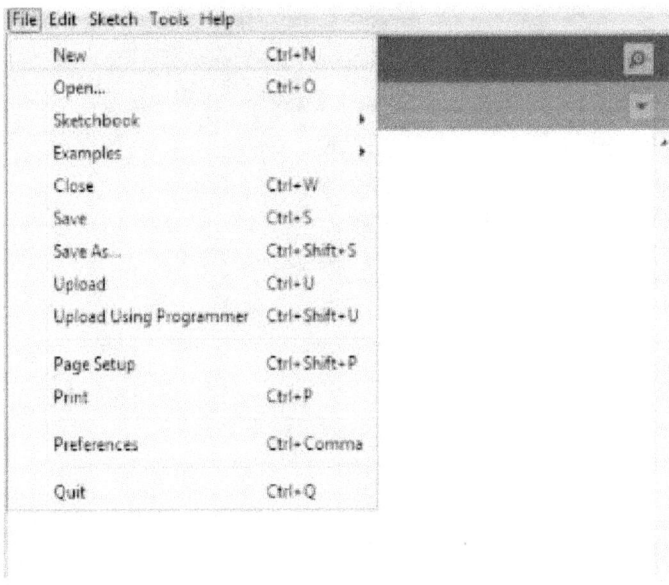

- The Arduino IDE provides a lot of completed examples

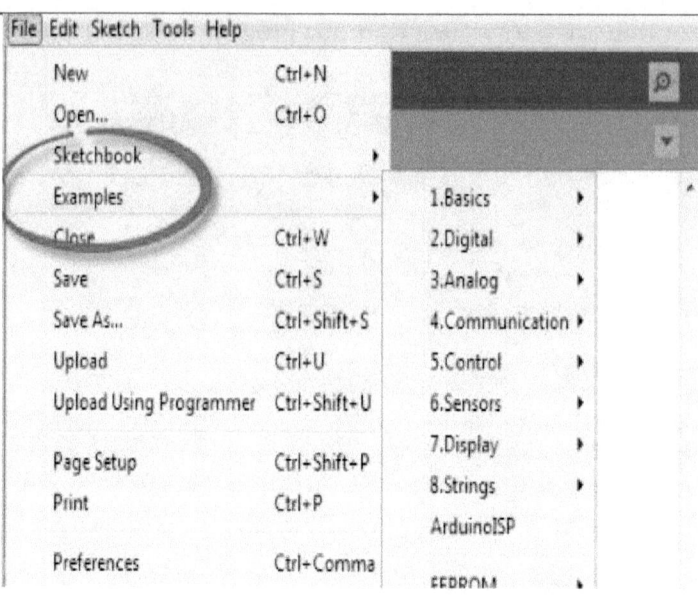

Questions

1. What is an Arduino?

2. Define the following:

- A resistor

- A digital multimeter

- An LED

Chapter 3

Getting started with Arduino

What you will learn in this chapter:

📇Write your first Arduino program

📇Understand the Arduino C language

📇Electronics basics

What you will need for this chapter:

📇An Arduino UNO board   📇Wires

📇Computer, or any type of PC   📇Breadboard

📇LEDs

📇Resistors   📇Push buttons

First example: LED blinking (Wiring)

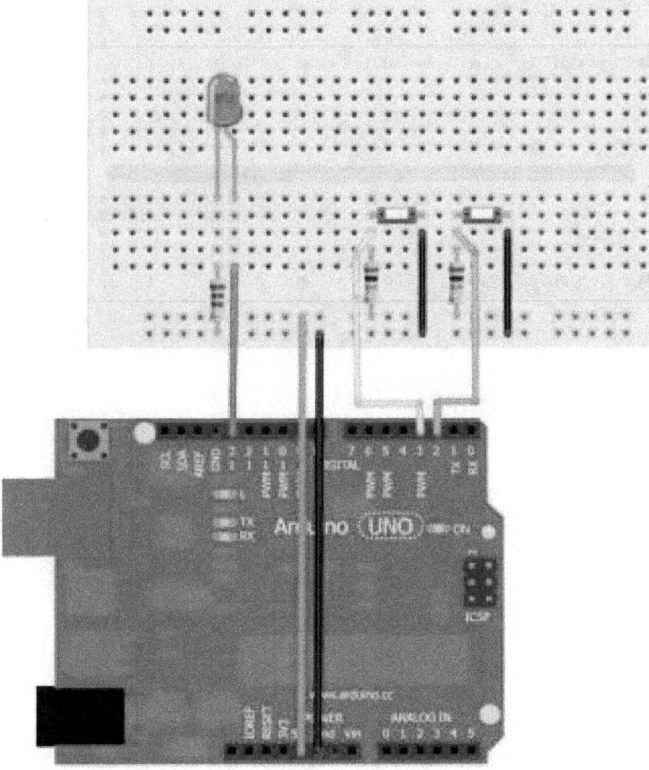

Steps

• Connect the longest leg (+) of the LED to pin number 13.

• Connect the other leg with the 560 ohm resistor.

- Connect the 5v pin and GND pin on the Arduino to the breadboard as shown.

First example: LED blinking (Coding)

Steps

- **Open the Arduino IDE, select File, then new.**

- **Write the following code:**

const int LED = 13;

void setup ( )

{

  pinMode(LED, OUTPUT);

}

void loop()

{

  digitalWrite(LED, HIGH);

  delay(1000);

digitalWrite(LED, LOW);

delay(1000);

}

After writing the code now you can press verify for the quick bar on the IDE and wait until "done compiling" then select upload to load the code on the Arduino board.

Congratulations! You have completed your first program on the **Arduino!**

Now it's time to explain the code

- Const int LED = 13;

This statement means that you will create a constant type integer with the name LED and value of 13.

- We use constants here to make it easy when naming (input and output) pins.

On the microcontroller in Example 1, we declare the constant to control the pin 13 by using name "LED" so we can use the name instead of using the pin 13 for readability.

- Void setup ()

{

   PinMode (LED, OUTPUT);

}

To set the pin number 13 to the output mode, this has the name "LED".

There are three steps to writing a program on an Arduino or any microcontroller.

- First of all, declaring the variables that we will use in the program.

- Secondly, it's really important to understand that for all of the digital pins we can set as an input or output pin. In our example we set the pin 13 as an output.

- PinMode (pin number, state)

 - PinMode-> function name.

- Pin number -> the number of the pin we will use.

- State - > to set the pin as an input or output.

- You should write "OUTPUT" or "INPUT" with capital letters.

- **Also**, we should write all of our input and output settings inside the function braces → {}

Setup () {Write you configuration here}, for example:

If we want to set pin 10, pin 11, or pin 13 as an output

and pin 2 as an input :

Void setup()

{

pinMode(10,OUTPUT);

pinMode(11,OUTPUT);

pinMode(13,OUTPUT);

pinMode(10,INPUT);

Note that every statement must end with semicolon";".

}

- The final step is to write what the microcontroller should do as in the first example :

Void loop()

{

digitalWrite(LED, HIGH);→ **turn on the LED.**

delay(1000); →**wait 1000 millisecond "1 second".**

digitalWrite(LED, LOW);→ **turn off the LED.**

delay(1000);→**wait 1000 millisecond "1 second"**.

}

- Keep in mind that the program will be written inside the function.

void loop () {Write your code here}

- digital write (LED, HIGH)

HIGH = 5 volt

LOW = 0 volt

First, we write the pin name and then the voltage.

- delay (1000)

To tell the microcontroller how much time to wait before the execution of the next instruction, we can control the time of the LED turning off or on.

- As we can see in our example:

digitalWrite(LED,HIGH);

delay(1000);

These instructions mean that the microcontroller will apply the 5v on the output pin which is connected to the LED. Then it will wait for 1,000 milliseconds.

Note that when we write programs on microcontrollers, time should be written in milliseconds instead of second "1000 millisecond = 1 second".

- There are two types of comments

// this is a one line comment

/* This is

a multiline

comment*/

- We use comments for readability or to describe the code and make it easy to understand.

- The Arduino IDE will ignore any comment.

Example 2 (Wiring)

In this example, we will use a pushbutton to turn the LED on or off.

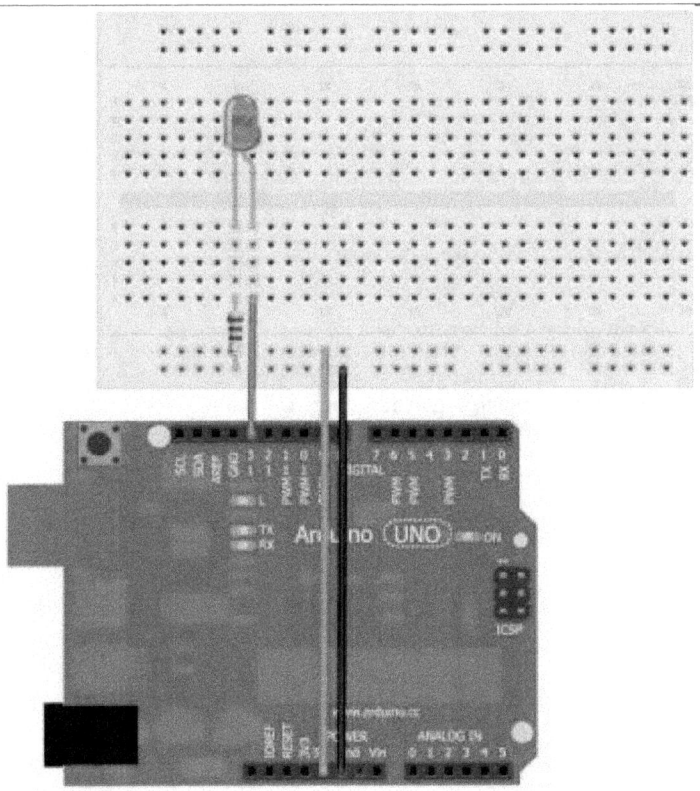

Parts you will need:

Breadboard

Push button

LEDs

10k ohm resistor

60 ohm resistor

Some wires

Steps:

- Put the push button on the breadboard.
- Connect one side of the button with the 5v using the wires.
- Connect the other side to the 10k ohm resistor.
- Connect the wire to pin 2 on the Arduino.
- Connect the other leg of the resistor to the ground.

Example 2 (Coding)

/*first step variables declaration and assignment */

const int ledPin = 13;

const int buttonPin = 2;

int val;

/*Second step define the pins and its directions */

void setup ( )

{

pinMode(ledPin, OUTPUT);

pinMode(buttonPin, INPUT);

}

/*Third step in writing the program */

void loop()

{

  val = digitalRead(buttonPin);

  if (val == HIGH)

    {

  digitalWrite(ledPin, HIGH);

  delay(1000);

  digitalWrite(ledPin, LOW);

  delay(1000);

    }

  else {digitalWrite(ledPin, LOW); } }

- Now click on the **verify** button and after the compilation, click the **upload** button to burn the code onto the Arduino board.

- Now it's time to explain the code.

- **In the first block**

    `int ledPin = 13;`

    `int buttonPin = 2;`

    `int val = 0;`

- We declared a variable called **ledPin** which was assigned to pin 13. Also, we declared another variable called **buttunPin** which was assigned to pin 2.

And we will use the Val variable to store the input state.

- **In the second block:**

    `void setup ()`

    `{`

```
pinMode(ledPin, OUTPUT);

pinMode(buttonPin, INPUT);

}
```

- We make the controller work with pin 13 as an output which was assigned before as "ledPin", then we set the pin 2 as an input to receive the digital signals.

Low or high

- **In the third block:**

Val = digitalRead (buttonPin);

In this line, the Arduino will measure the voltage and store the value in the variable Val using the digitalRead () function. For example:

- f the button was pressed, so the value will be 5v = HIGH.

- therwise the value will be 0v = LOW.

```
If (Val == HIGH)

{
```

```
digitalWrite(ledPin, HIGH);

delay(1000);

digitalWrite (ledPin, LOW);

}

else

{

digitalWrite (ledPin, LOW);

}
```

In the above code we used if / else statement to compare variables and make the microcontroller do some actions based on the results.

- The Arduino will measure the voltage and store the value in Val.

- If the value is equal to 5v or higher, the controller will turn on the LED for 1 second and turn it off for 1 second.

- Unless the value isn't equal to 5, so the microcontroller will not turn on the LED and will be off.

Example 3: (Wiring) LED blinking using two push buttons.

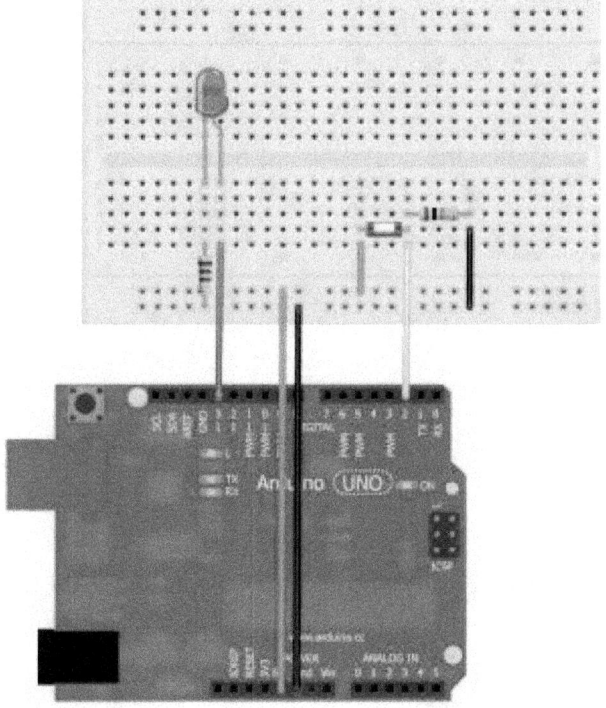

Parts you will need:

- rduino Uno

- readboard

- ED

- 0 k ohm resistors (2)

- ush buttons (2)

- 60 ohm resistor

- ires

In this example, we will apply what we have learned in the last 2 examples.

Example 3 (Coding)

- From the Arduino IDE select file > new, and write the following code:

/* declaration and assignment of the variables*/

Const int ledPin = 13;

Const int inputPin1 = 2;

Const int inputPin2 = 3;

/* define the direction of the pins */

Void setup ()

{

　pinMode (ledPin, OUTPUT);

　pinMode (inputPin1, INPUT);

　pinMode (inputPin2, INPUT);

}

/* the main program */

Void loop ()

{

　If (digitalRead(inputPin1) == HIGH)

　{

```
digitalWrite (ledPin, LOW);
}
else if (digitalRead (inputPin2)== HIGH);
{
digitalWrite (ledPin, HIGH);
}
}
```

- In this example, we used **else if** for adding more than one condition in one if statement.

Chapter 3 Review

Void setup () → This function is used to set the pins' direction as an input or output.

Void loop () → In this function body, you will write your main program.

Int name = value; A statement to define a variable and its value.

example: const int LED = 13; statement to define a constant.

`pinMode(pin number, state);` To define the pin number and its direction.

`example: pinMode(11,INPUT);`

`digitalWrite(pin number, state)` To determine the voltage on the pin.

`example: digitalWrite(13, OUTPUT);`

`digitalRead(pin number)` To read the voltage from the pin.

`example: digitalRead(4);`

`delay(time)` This function used to determine how much time the Arduino should wait.

`example: delay(1000);`

`if (the condition) {what to do}`

`else if (another condition)`

`{what to do }`

`else(last condition)`

{what to do}

Conditional statements to determine what to do based on some variables

Data Type	example	Value (range)
Integer	int LED = 13;	From -32768 to 32768
Float	float sensor = 12.5;	With decimal numbers
Character	char name = 'a';	character/text
Long	Long variable = 99999.9;	From -2,147,483,648 to 2,147,483,648
Byte	Byte number = 55;	from 0 to 255

Questions

1. Write a code to blink an LEDLED 30 times in 1 minute.

2. Write a code to blink two LEDs in the reverse way.

3. Design the circuit of the second example using any tool like fritzing.

4. How many bits are in one byte?

5. Extend the code and the circuit of the second example using push buttons.

Chapter 4

Inputs, Outputs, and Sensors

What you will learn in this chapter:

📖 Introduction to signals

📖 Work with sensors

📖 Understand PWM

What you will need for this chapter:

▦ Arduino UNO board   ▦ Multimeter

▦ Sensors   ▦ Resistors

There are two types of signals:

- Digital signal:

A digital signal refers to an electrical signal converted into bits (0s / 1s).

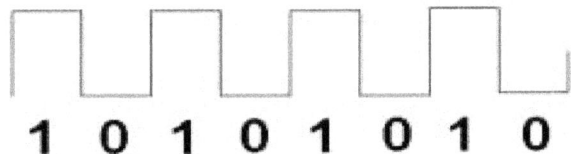

- Analog signal:

Unlike a digital signal, an analog signal is any continuous signal for which the time varying feature of the signal is a representation of some other time varying quantity.

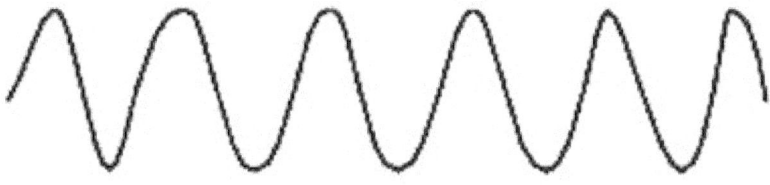

Why are analog signals important?

Analog inputs like the voltage of some sensors are a result of changing some factors. For example:

Photo – resistor: which is an electrical resistor that changes its value depending on the amount of light.

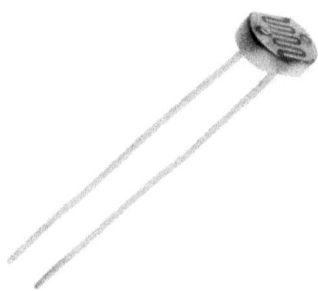

We can measure the voltage on this resistor using the multimeter.

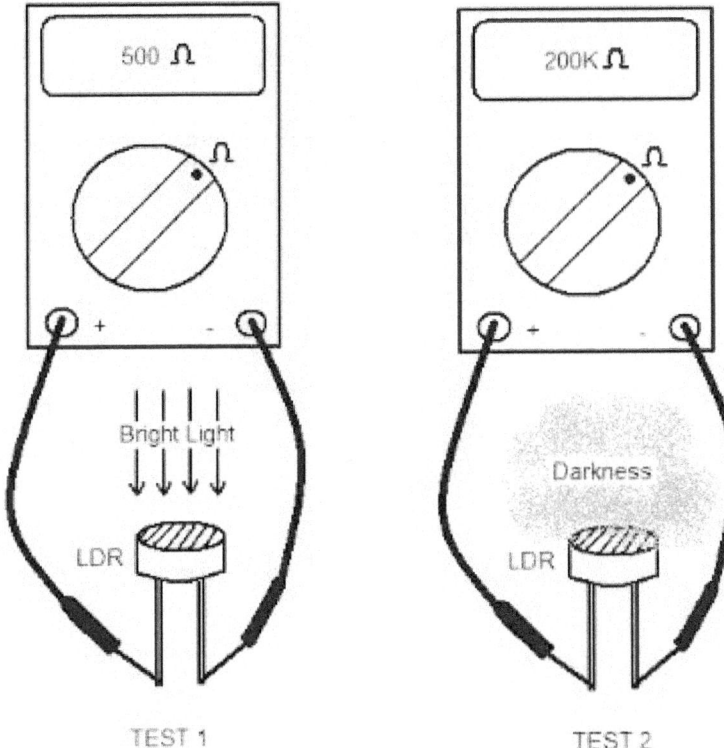

TEST 1   TEST 2

- We can use this phenomenon to measure any other environmental factor using proper sensors that convert the factor into analog signals such as light, temperature, humidity, power, etc.

- On the Arduino UNO (ATMega 328p), there are six input pins for the analog signals it start from A0 to A5, and it can measure voltages with 4.8

millivolt, and that means it's very accurate when measuring a lot of

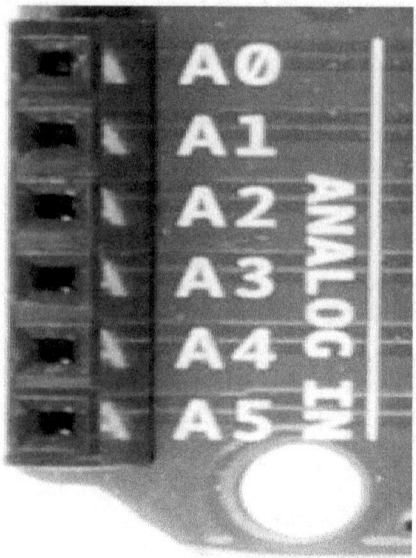

applications.

• In this chapter, we will work with analog sensors like the photo resistor and the temperature sensor. The TMP35 or LM35 is actually a simple transistor that changes it's the voltage by the changing of the temperature.

How do sensors generate analog signals?
Let's take the temperature sensor as an example: the temperature sensor contains a very sensitive transistor which is made from silicon. And as we know, silicon is highly affected by the temperature.

The temperature sensor has the following:

1. Input **Vin** (2.2v to 5.5v).

2. Signal leg **Vout** to get the measurement.

3. The ground leg **GND** to connect it with any ground point.

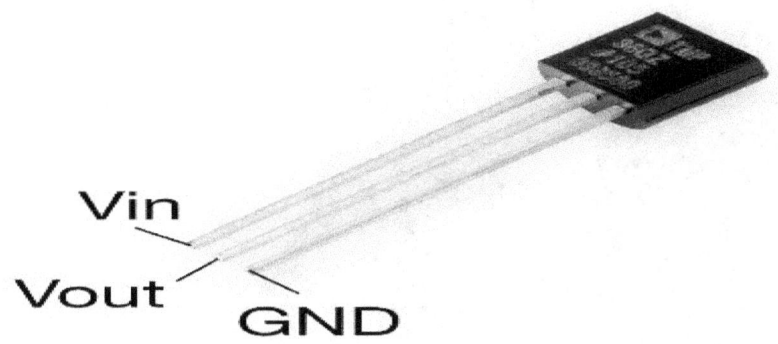

Components you will need for this example:

- Multimeter

- AAA 1.5v battery (2)

- Temperature sensor (TMP35 or TMP35 or LM35)

Steps

- Bring the two AAA batteries and put them together in the battery holder so you will get 3 volts.

- Connect the red wire with that of the battery holder to the temperature Vin leg.

- Connect the black wire of the battery holder to the temperature sensor GND leg.

- Put your multimeter to the voltage mode as shown below:

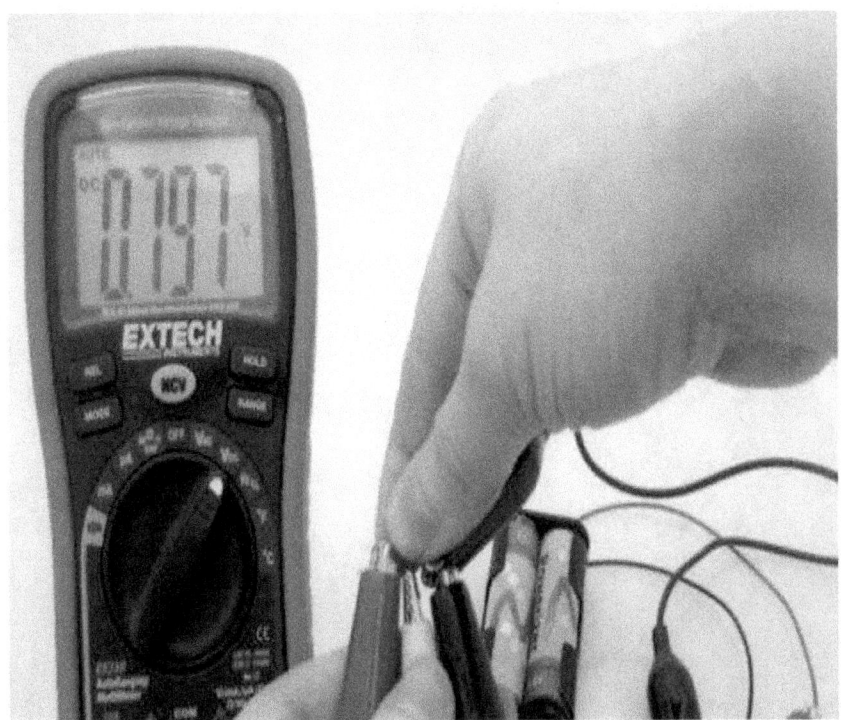

- Connect the GND leg to the black probe, and connect the red probe to the Vin leg as shown.

- Note the reading of the voltage on the multimeter. It should be 0.76 volts.

- Now put your hand on the sensor (this movement will raise the temperature) and the note the reading of the multimeter.

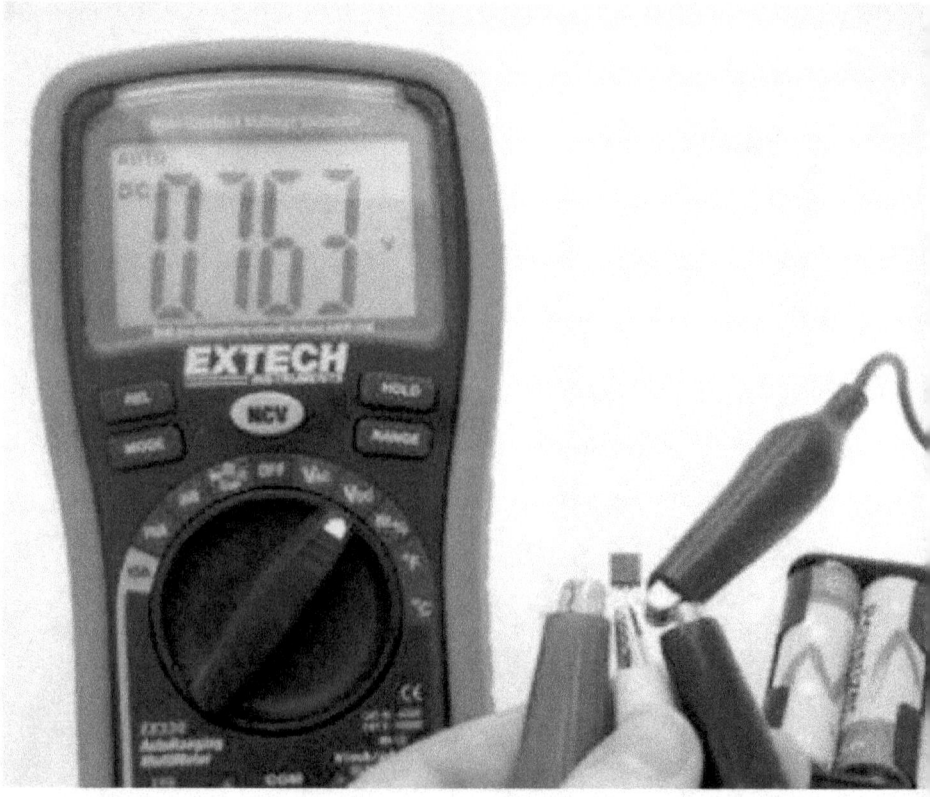

You will find that the reading becomes higher.

- As with any sensor work in the same manner of the temperature sensor, it behaves depending on the environmental factor and changes its internal resistor, so it changes the output voltage which can be measured by an analog sensor.

Example 4: Control light amount using a potentiometer (wiring)
- In this example we will use a potentiometer to get a changeable voltage (analog input) and we will turn on/off the LED depending on the value of the analog input.

Example components

- Arduino UNO board
- Breadboard
- LED
- 560 ohm resistor
- 10 k ohm potentiometer
- Wires

Connect the components as shown:

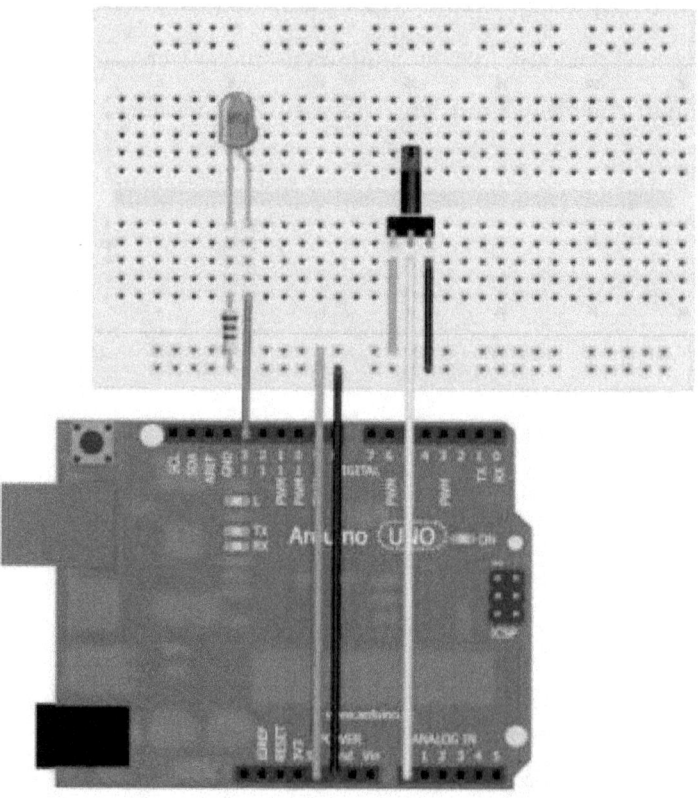

Example 4: Control light amount using potentiometer (Coding)

//create new file form the Arduino IDE and write the following code:

```
const int sensorPin = A0;

const int LedPin = 13;

int sensorValue;


void setup ()

{

 PinMode (LedPin, OUTPUT);

}


void loop()

{

  sensorValue = analogRead(sensorPin);

digitalWrite(LedPin, HIGH);

delay(sensorValue);
```

digitalWrite(LedPin, LOW);

delay(sensorValue);

}

In this example, we use one of the most important functions in the Arduino language; **analogRead(pin number).** This function reads the voltage as an analog signal (the microcontroller can measure voltage from 4.8 millivolt to 5 volt), and it also converts these values to digital values from 0 to 1,024. This conversion is called **analog to digital converting (ADC).**

For example:

If the input voltage to the A0 equals the following values:

4.8millivolt = 1 in digital

49millivolt = 10 in digital

480millivot = 100 in digital

1volt = 208.33 in digital

2volt = 416.66 in digital

5volt = 1024 in digital

`sensorValue = analogRead(sensorPin);`

- In this statement, the microcontroller will store the value of the sensor reading in the sensor value variable, and then the microcontroller will turn on/off the LED for a period of time equal to this variable (sensorValue).

- In this example we have used a variable resistor, so we could change the its value of the resistance.

Example 5 photo resistor as light sensor (Components)

- Arduino UNO board

- Breadboard

- LED

- 560 ohm resistor

- Photo resistor

- wires

Example 5: Photo resistor as light sensor (Wiring)

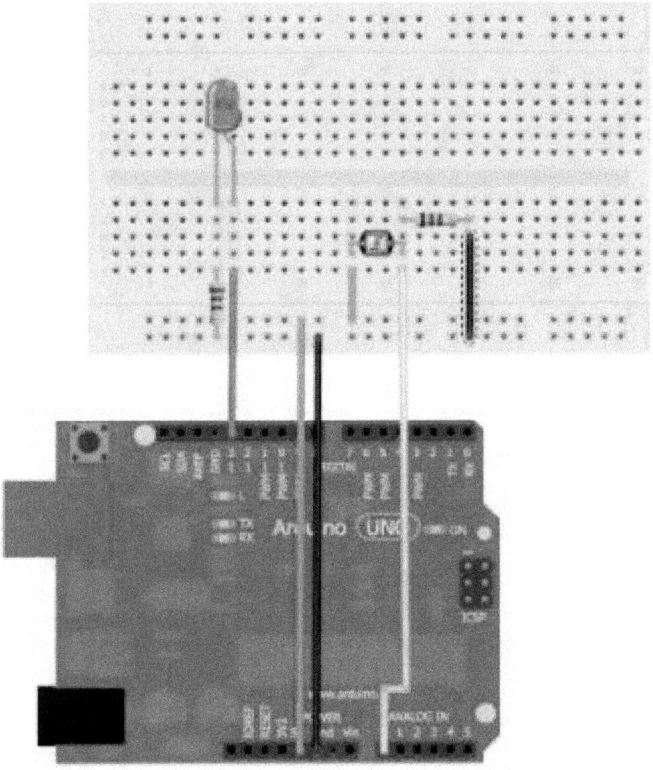

• Connect the components as shown:

Example 5: Photo resistor as light sensor (Coding)

// select new file from the Arduino IDE

const int lightPin = A0;

const int ledPin = 9;

```
int lightLevel;

void setup ()
{
pinMode(ledPin, OUTPUT);
}

void loop ()
{
  lightLevel = analogRead(lightPin);
  lightLevel = map(lightLevel, 0, 900, 0 , 255)
  lightLevel = constrain(lightLevel, 0, 255);
  analogWrite(ledPin, lightLevel);
}
```

- **Now** you can upload this code on your Arduino board and look what will happen to the LED after focusing the light on the photo resistor. Then put your hand on the photo resistor and look what will happen to the LED.

- analogWrite(pin number, value);

This function generates an analog output, and this function can be applied to all of the pins with pulse width modulation (PWM) .

They are pin 3, pin 5, pin 6, pin 9, pin 10, and pin 11 (any pin with ~ **sign**).

What is pulse width modulation?

Pulse width modulation is a technique for getting analog results by digital means. Digital control is used to create a square wave, a signal switched between on and off. This on-off pattern can simulate voltages in time that the signal spends off. The duration of "on time" is called the pulse width. To get varying analog values, you can change or modulate that pulse width. If you repeat this on-off pattern fast enough with an LED, you can generate voltage between 0v and 5v using digital values as shown earlier.

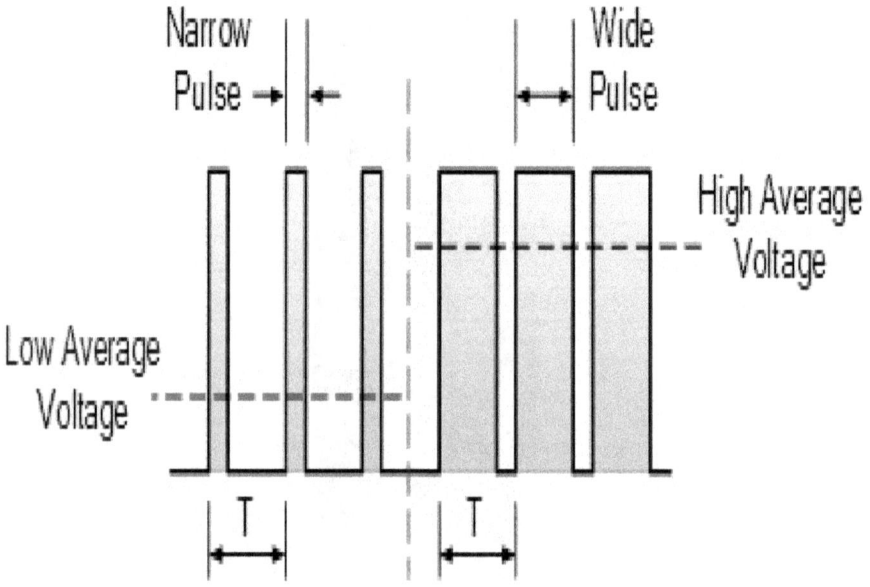

How we can use it?

A lot of electric components are dealing with different voltage values.

For example, when you apply 3 volts to the LED you will get a very small amount of light, and if you raise the voltage to 4 volts you will find out that the light will be more bright and so on.

And if you use a motor, for example, when you increase the voltage the speed of the motor will be faster.

Example 6: LED with PWM (wiring)

- **Connect the components as shown:**

Example 6: LED with PWM (coding)

// open the Arduino IDE and select new file then write the following code:

```
const int ledPin = 11;

int i = 0;

void setup( )

{

pinMode(ledPin, OUTPUT);

}

void loop()

{

for (i = 0; i < 255; i++) // LED will be lighter

{

analogWrite(ledPin, i);

delay(10);

}

for (i = 255; i > 0; i--)    //LED will be darker

{
```

analogWrite(ledPin, i);

delay(10);

}

}

for (i = 0; i < 255; i++)

- In the previous example, we used a new statement which is the for loop statement. You can use the for loop if you want to run the same code over and over again, each time with a different value.

I = 0 → the initial value

I < 255 → to set your condition

I++ → is the iterator in this example will add 1

I++ → I = I +1

Questions

1. Describe the difference between digital and analog signals.

2. What is pulse width modulation?

3. Design a circuit to turn on/off five LEDs in sequential order.

4. Write the code for Example 3.

Chapter 5

Computer interfacing with an Arduino

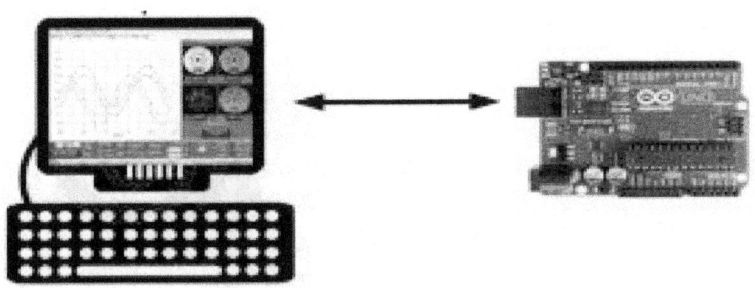

What you will learn in this chapter:

📇 How to connect your Arduino with your computer

What you will need for this chapter:

📇 An Arduino UNO board

📇 Breadboard

📇 Sensors

📇 Wires

FTDI Chips

- All of the Arduino boards have the capability of sending and receiving data to and from the computer directly through the USB port except the Mini and Lilypad Arduino boards. But you can also connect these boards with the computer using the FTDI interface, which is a small chip used to exchange the data between the Arduino or any microcontroller and the computer.

- **In the last examples** we used the Arduino to read some sensor values, like light and temperature, to show the results on the LED.

- **In this chapter,** the serial interface will send the sensor values to the computer, and we can get the calculations easily.

Example 7: Temperature sensors with serial interface (Components)
- An Arduino UNO board

- Breadboard

- Temperature sensor (TMP 36 or LM35)

- A – B USB cable

Example 7: Temperature sensor with serial interface (Wiring)

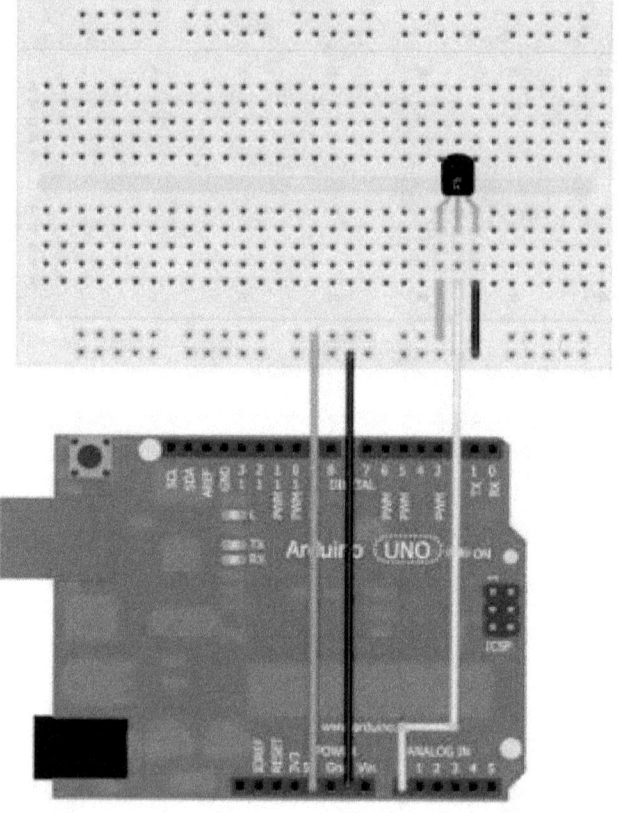

Example 7: Temperature sensor with serial interface (Coding)

const int sensorPin = A0;

int reading;

float voltage;

float temperatureC;

void setup( )

{ Serial.begin(9600); }

void loop ( )

{

reading = analogRead(sensorPin);

voltage = reading * 5.0/1024;

Serial.print (voltage);

Serial.println(" volts");

temperatureC = (voltage - 0.5) * 100 ;

Serial.println("Temperature is: ");

Serial.print(temperatureC);

Serial.println(" degrees C");

delay(1000);

}

● **After verifying and uploading** the code, click on the Serial Monitor as shown:

```
File Edit Sketch Tools Help

                                        Serial Monitor

sketch_jun18a §
int sensorPin = A0;
int reading;
float voltage;
float temperatureC;

void setup( )
{ Serial.begin(9600); }

 void loop ( )
 {
reading = analogRead(sensorPin);
voltage = reading * 5.0;
voltage /= 1024.0;
Serial.print(voltage);
Serial.println(" volts");
temperatureC = (voltage - 0.5) * 100 ;
Serial.print(temperatureC); Serial.println(" degrees C");
delay(1000);
 }
```

- You will see this menu that shows the temperature sensor readings.

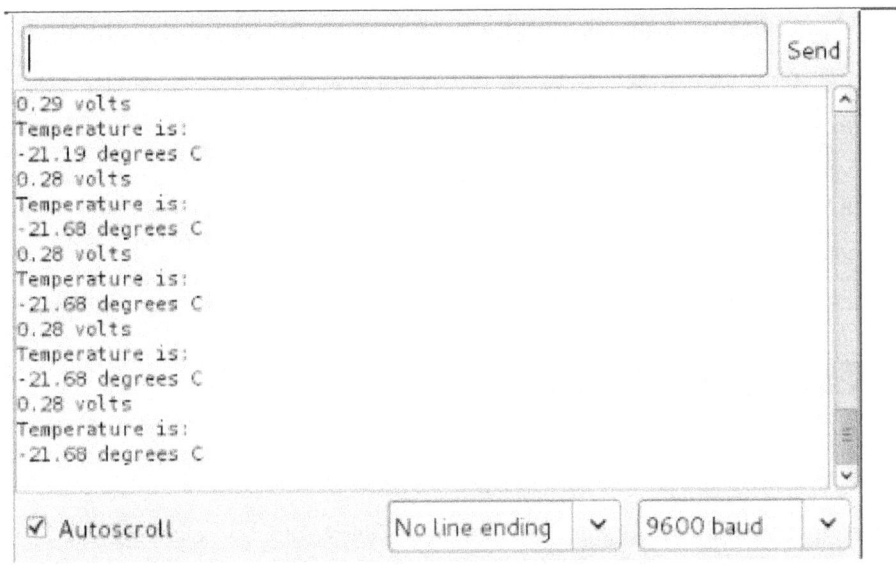

- **Now** try to raise the temperature using any heat source.

- You should be aware that this sensor can handle 150 Celsius.

- (-) This symbol doesn't mean negative, but it is a temporary programming error.

Example7: Temperature sensor with serial interface (Explanation)

Serial.begin (9600);

- We write this statement to start the communication between the Arduino and the computer through the USB port, so we can receive and send data to and from the computer.

- There are two variables in our code (voltage, TemperatureC) that have been defined with float instead of int because the temperature sensor is a very accurate sensor , and the result will be in floating points number not integers.

`reading = analogRead(sensorPin);`

- This instruction is used to record the analog input in the A0 pin.

As we mentioned before that the microcontroller converts the analog signal into digital values from zero to 1024, we used this instruction:

`voltage = reading * 5/1024;`

- After the conversion of digital values to voltage, we used **Serial.print (voltage);**

to send this value to the computer and show it on the Arduino IDE.

- **Serial.print ("voltage");** This instruction is used to print the word "voltage"

after its value.

- **TemperatureC = (voltage – 0.5) *100;** This instruction is to convert the voltage values to temperature degrees in Celsius, and print the value then the word "Temperature" and "degree C".

Serial.print(TemperatureC);

Serial.println("degree C");

● The last line of code is **delay (1000);** to make the microcontroller wait one second before sending the voltage and the temperature value to the computer again.

Example 8: Showing the strength of the LED light on the serial monitor (Wiring)

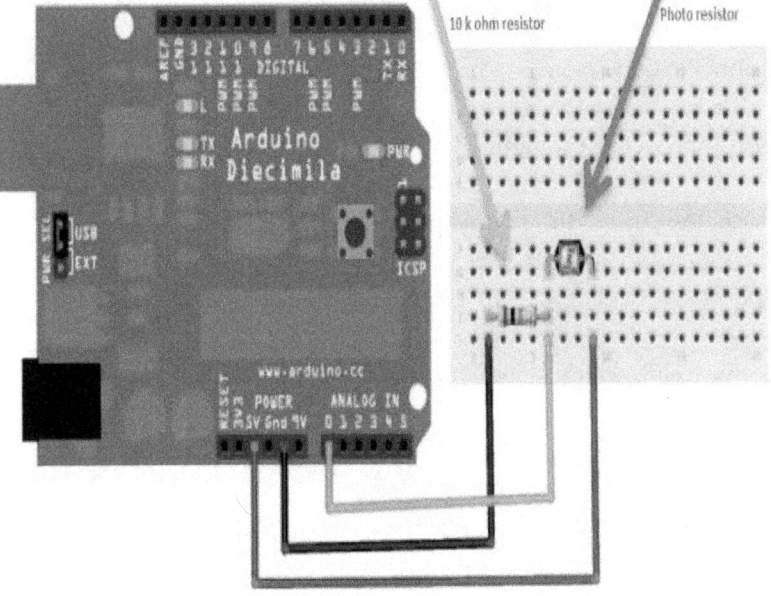

Example 8: Showing the strength of the LED light on the serial monitor (Coding)

```
const int photocellPin = A0;

int photocellReading;

void setup(void)

{ Serial.begin(9600); }

void loop(void)

{

photocellReading = analogRead(photocellPin);

Serial.print("Analog reading = ");

Serial.print(photocellReading);

if (photocellReading < 10) { Serial.println(" - Dark");}

else if (photocellReading < 200) { Serial.println(" - Dim");}

else if (photocellReading < 500) {Serial.println(" - Light"); }

else if (photocellReading < 800) { Serial.println(" - Bright"); }

else {Serial.println(" - Very bright"); }

delay(1000);

}
```

After uploading the code on the Arduino, click on the serial monitor.

```
int photocellPin = A0;
int photocellReading;
void setup(void)
{ Serial.begin(9600); }

void loop(void)
{
  photocellReading = analogRead(photocellPin);
  Serial.print("Analog reading = ");
  Serial.print(photocellReading);

  if (photocellReading < 10) { Serial.println(" - Dark");}
  else if (photocellReading < 200) { Serial.println(" - Dim");}
  else if (photocellReading < 500) {Serial.println(" - Light"); }
  else if (photocellReading < 800) { Serial.println(" - Bright"); }
  else Serial.println(" - Very bright"); }
  delay(1000);
}
```

- **Now** try to do the following:

- Focus the light on the photo resistor

- Cover the photo resistor with any transparent piece of clothing

- Cover the photo resistor with your hand and make sure no light is on it

• This is what you will see:

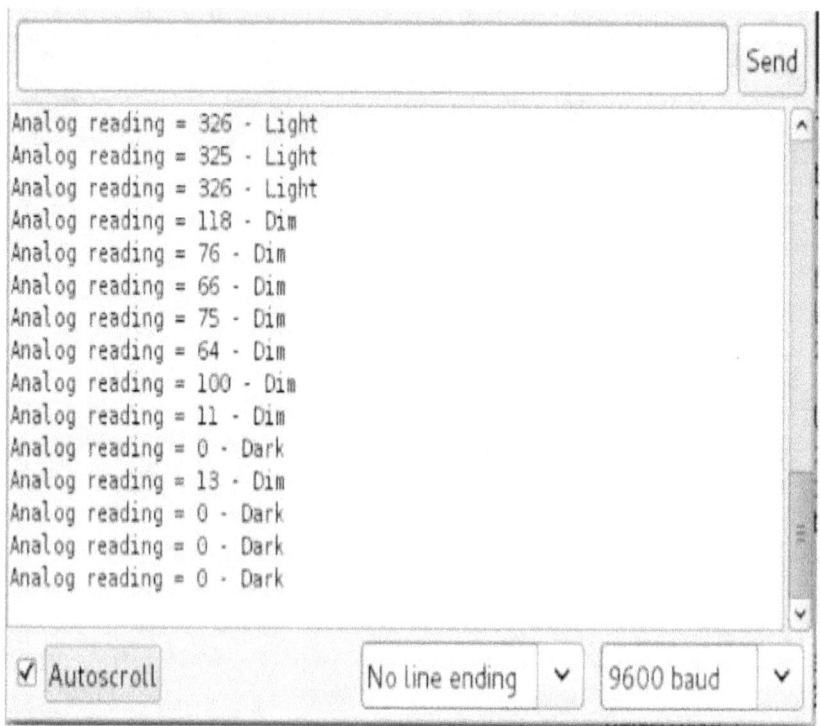

• **Dim** → the amount of light will be small

• **Dark** → there is no light

• **Light**→ there is a moderate amount of light

• **Bright light**→ the brightness of the light is very high

Example 9: Turn you LED on/off using your computer (Components)

- An Arduino UNO board

- Breadboard

- LED

- 560 ohm resistor

- Wires

- **In this example** will use the computer to control the LED instead of using a switch, and the Arduino will receive the command using the serial monitor through the USB port.

Example 9: Turn your LED on/off using your computer (Wiring)

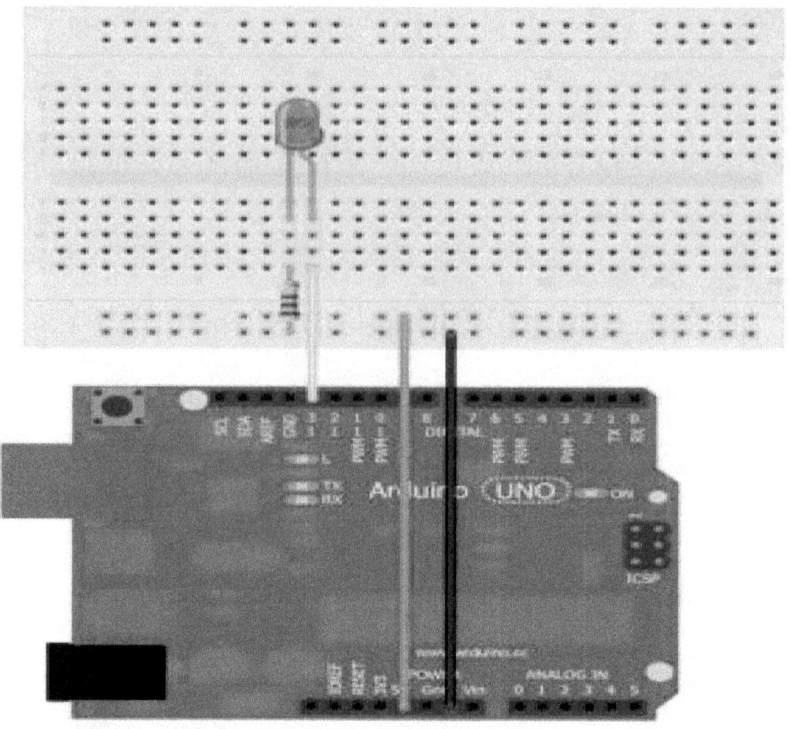

Example: 9 turn on / off your LED using your computer (Coding)

int ledPin=13;

int value;

void setup ()

{

Serial.begin(9600);

```
pinMode(ledPin,OUTPUT);

}

void loop ()

{

value = Serial.read();

if (value == '1') {digitalWrite(ledPin,HIGH);}

else if (value == '0') {digitalWrite(ledPin,LOW);}

}
```

After the uploading of the code on the Arduino, click on the serial monitor icon and you'll find a search bar. Write "1" on it, and click send. Then write "0", and watch what will happen to the LED.

- In this example, we have used the **Serial.Read();** instruction to read the data that was sent from the computer to the Arduino though USB, also we added the variable "value" to store the data.

Then we used the if else statement.

- if value == 1 the microcontroller will turn on the LED

- if value == 0 the microcontroller will turn off the LED

Questions

1.
ow do you can make the Arduino communicate with the computer?

2.
hat is the FTDI Chip, and how can you use it?

3.
esign a circuit to connect the Arduino with a temperature sensor and an LED.

4.
rite the code for Example 3 and control the LED based on the readings of the temperature sensor.

Chapter 6

The Motors

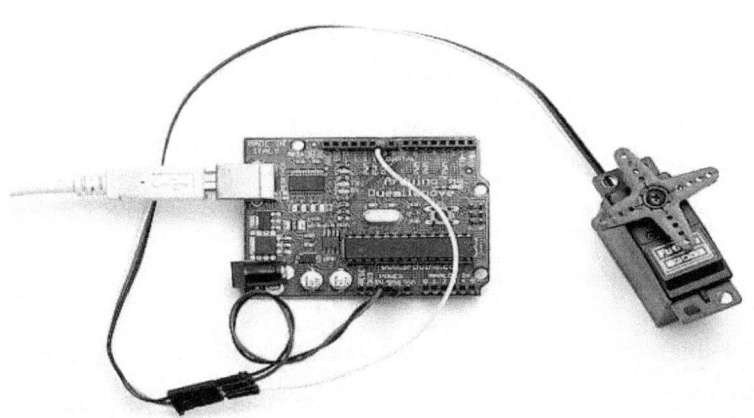

What you will learn in this chapter:

- Introduction to motors

- Types of the motors

- Interfacing the motors with the Arduino

What you will need for this chapter:

- An Arduino UNO board

- Breadboard

- Transistors

- DC/Servo motor

Intro

• The motor is a very important electrical component that you will need in a lot of projects because it's the element that converts electrical energy into mechanical energy.

• You can find the motors in a lot of applications such as robots, CD drives, toys, etc.

• There are mainly two types of motors:

- Direct current motors (DC – Servo - Stepper)

- Alternative current motors (1Phase – 3Phase)

• In this chapter, we will use the first type which are the direct current motors (DC – Servo - Stepper), especially the DC and the Servo motor

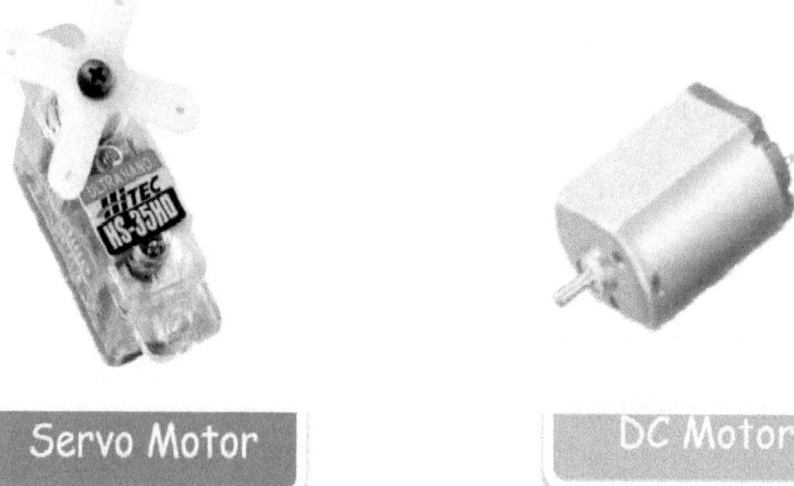

Servo Motor   DC Motor

Example 10: Using the direct current motor "DC Motor" (Components)

- An Arduino UNO Board

- Breadboard

- DC motor

- 2N2222 or PN2222 Transistor

- 1N4001 Diode, or any alternative

- 2.2 k ohm resistor

- Some wires

- USB cable

- **In this example** we will use the small size of the direct current motor that is usually used in toys. It can work with 3volts to 9volts, and you can easily find this type of motor in an electrical components store or any toy store.

Example 10: Using the direct current motor "DC Motor" (Wiring)

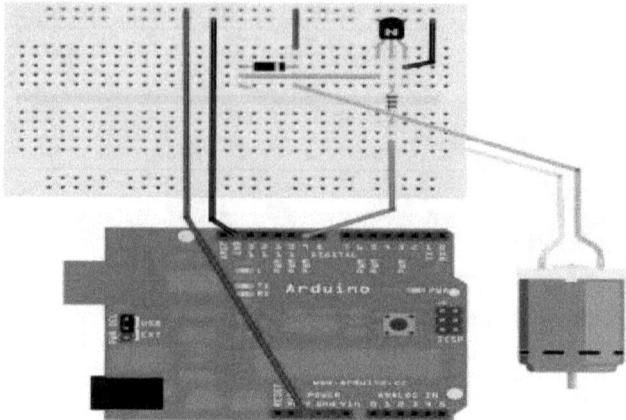

• After connecting the components with the Arduino UNO, you can start writing the following code and upload it on the Arduino:

int motorPin = 9 ;

int onTime = 2500 ;

int offTime = 1000 ;

void setup) (

{pinMode(motorPin, OUTPUT); }

void loop) (

{

analogWrite(motorPin,100);

delay(onTime);

digitalWrite(motorPin, LOW);

delay(offTime);

analogWrite(motorPin,190);

delay(onTime);

digitalWrite(motorPin, LOW);

delay(offTime);

analogWrite(motorPin,255);

delay(onTime);

digitalWrite(motorPin, LOW);

delay(offTime);

}

Example 11: Using the direct current motor "Servo Motor" (Components)

- An Arduino UNO board

- Breadboard

- DC motor

- 2N2222 or PN2222 Transistor

- 1N4001 Diode or any alternative

- 2.2 k ohm resistor

- Some wires

- USB cable

Example 11: Using the direct current motor "Servo Motor" (Wiring)

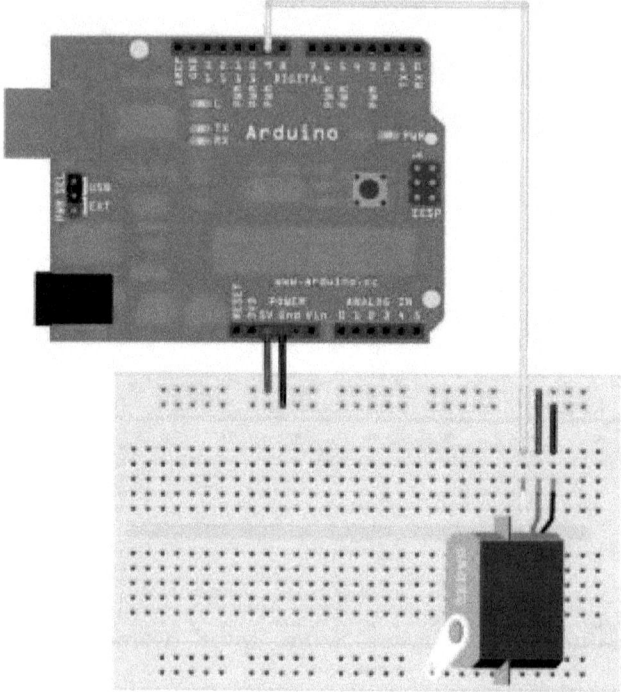

Example: 11 using the direct current motor "Servo Motor" (Coding)

#include <Servo.h> // additional library for the servo motor

Servo myservo;

int pos = 0;

```
void setup()

{

myservo.attach(9);

}

void loop()

{

for(pos = 0; pos < 180; pos += 1)

{

myservo.write(pos);

delay(15);

}

for(pos = 180; pos>=1; pos-=1)

{

myservo.write(pos);

delay(15);

}}
```

- After connecting the components to the Arduino, you can upload the code on the board.

- Also, you can find a sample of the code in the completed examples from the Arduino IDE.

Examples – servo – sweep

Questions

1. Describe the different types of motors.

2. Design a circuit to connect servo motors and Arduino.

3. Extend Example 3 using a LED, and make the following:

- If the motor is on turn on the LED

- If the motor is off turn off the LED

4. Using a DC motor and a LED, do the following:

- If the speed of the DC motor is the max → make the LED flashing quickly

- If the speed of the DC motor is a moderate speed → make the LED flashing in a moderate speed

- If the speed of the DC motor is a low speed → make the LED flashing in a low speed

Chapter 7

Advanced Inputs and Outputs

What you will learn in this chapter:

📇 Learn the different types of displays

📇 Understand relays

What you will need for this chapter:

📇 An Arduino UNO board

Different types of displays (LCD, keypad...)

Intro

- **In the previous chapters**, we used simple inputs and output devices with Arduino such as switches, LEDs, etc.

- **In this chapter**, I will show how to use the advanced inputs and outputs like:

• Liquid Crystal Display "LCD"

• The keypad

• The LED matrix

• The relays

- Let's take a look at the liquid crystal display, or LCD. This type of display was made from crystal, and we will use the most two popular types which are:

- Character liquid crystal display "LCD"

- Graphical liquid crystal display "GLCD"

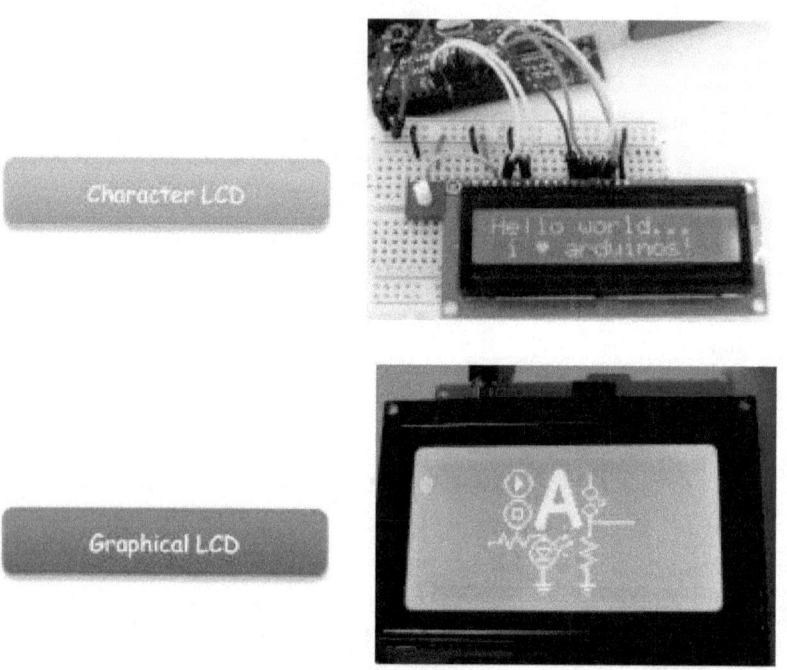

- The character LCD has the capability of displaying characters, numbers, or symbols (like the characters we type from the keyboard). You can find different sizes and colors of the character LCDs.

Green 16x2 LCD

Blue 16x2 LCD

Green 20x4 LCD

- As an example, 16x2 means the following:

- Number of the lines is two

- Number of the characters in every line is 16

You can choose from different sizes and colors.

Example 10: 16x2 LCD interfacing (components)

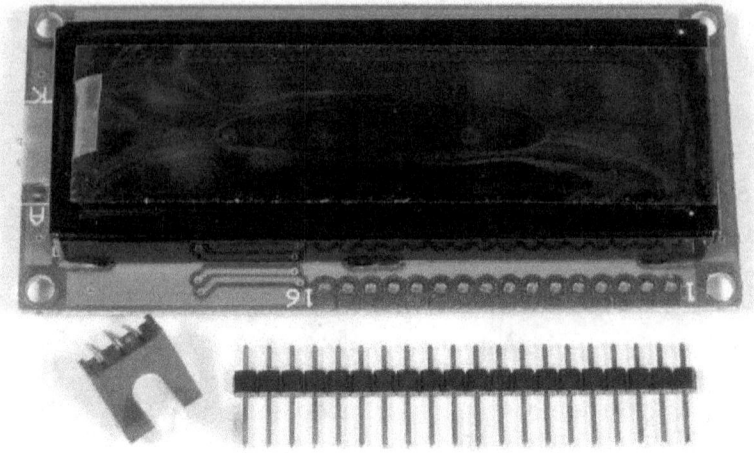

- 16x2 LCD with blue light (or any color you prefer)

- Copper pin headers 16 points

- 10 k ohm potentiometer

- Soldering iron

- Soldering wire

Steps
- Solder the pin headers with the LCD points using the soldering iron.

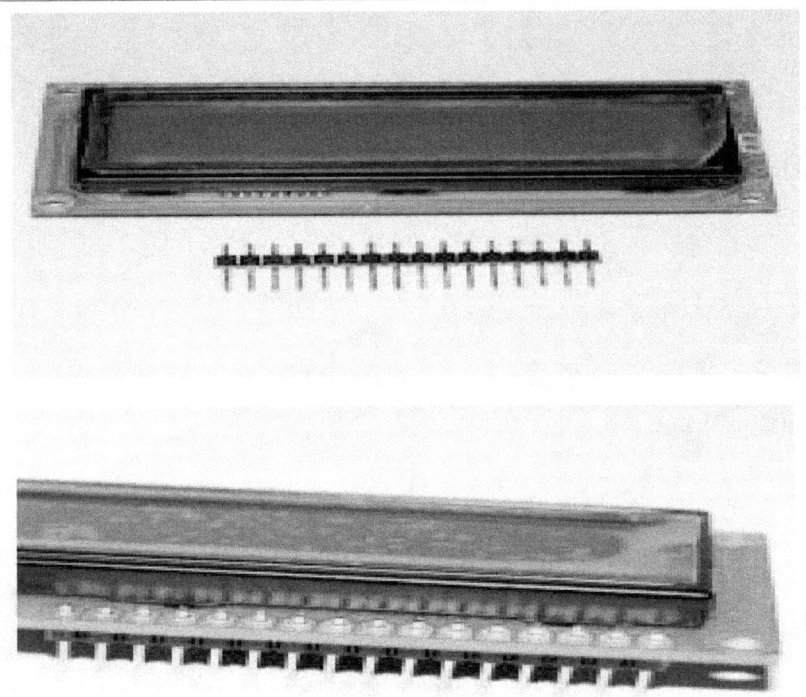

- Solder the first point using the wire and the soldering iron, and wait for 5 seconds to make sure that the point is not too hot, also don't try to touch it with your hand.

- Solder the last point and hold the LCD from the both sides.

- Repeat all of the procedures again for all 16 pins as shown:

- After finishing the soldering, it's now time to put it on the breadboard.

- Connect the Arduino 5v pin with red line pins on the breadboard, and the GND pin with the blue line pins on the breadboard.

- Connect pin number 16 on the LCD to the GND line, and pin number 15 to the positive line as shown in this picture:

- Connect the Arduino with the USB cable or the battery, and look at the light on the LCD.

- The color of display may be different depending on your LCD color choice. There are also other colors like red, white, green, and blue.

Use the potentiometer to control the brightness of the display.

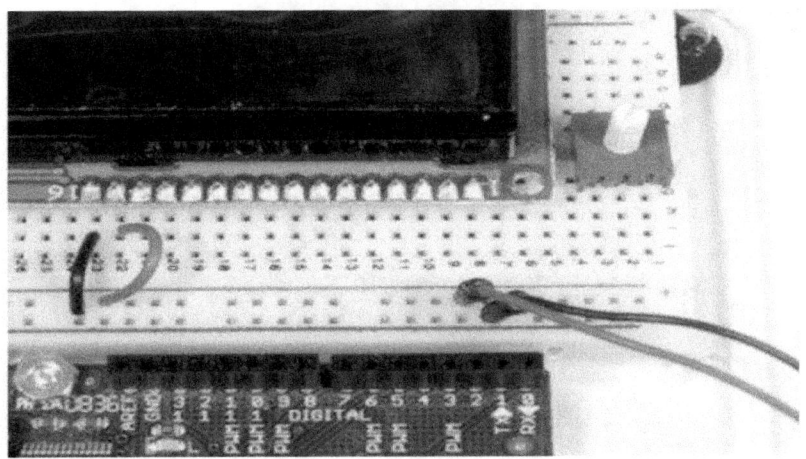

- This is an optional step that you can skip; the goal of using the potentiometer is to control the amount of current inside the LCD, so we can control the brightness on your display.

- Connect one of the legs of the potentiometer to the red positive line, and the other leg with the black negative line.

- Connect the middle leg of the potentiometer to the third pin on the LCD as shown.

- Now connect pin number 1 to the ground, and pin number 2 to the positive line on the breadboard as shown below:

- Connect the battery to your board, and rotate the hook of the potentiometer, and watch the difference on the light brightness of your display.

- The goal of all previous steps is to connect the LCD to the potentiometer and the battery to control the brightness of the display.

- In some projects we may use pin number 5, which is called RW, but in our project we will connect it to the ground.

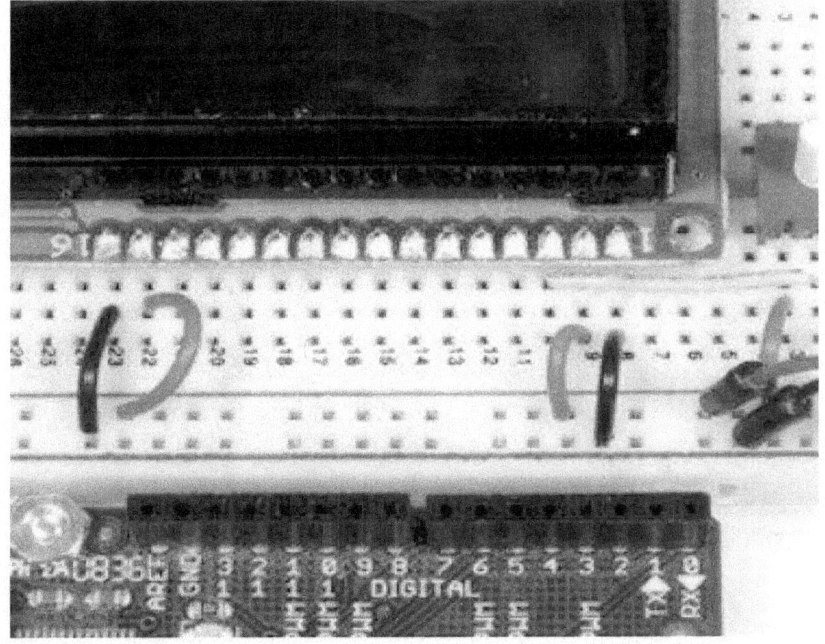

- Now connect pin number 4 on your LCD to pin number 7 on the Arduino.

- After that, you can connect pin number 6 on your LCD to pin number 8 on the Arduino board as shown:

- Connect pin number 14 on your LCD to pin number 12 on your Arduino board.

- The final step is to connect pins number 11, 12, and 13 on your LCD to pins number 10 and 11 in the same order as in the following picture:

- This is the final circuit for connecting the Arduino to the LCD.

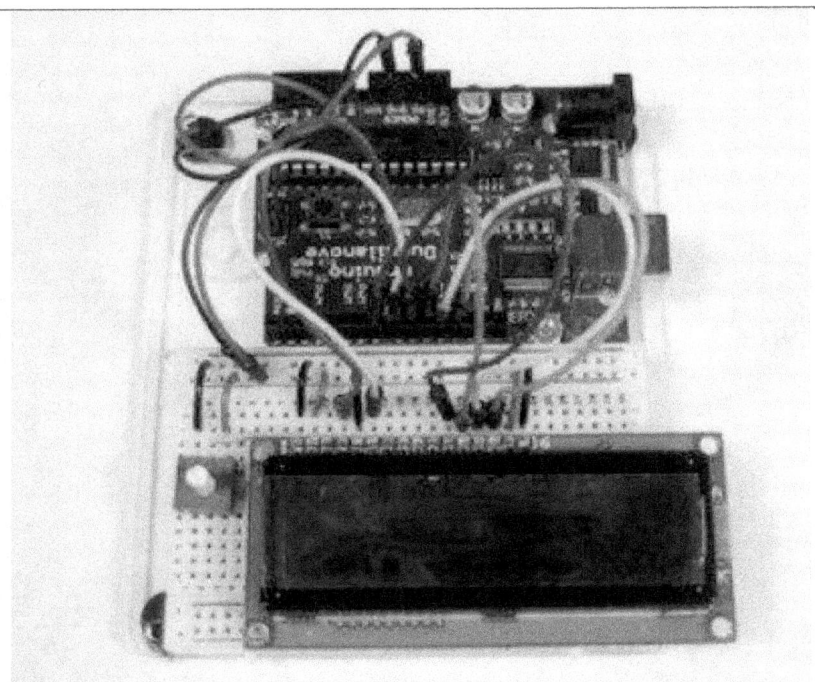

- Now it's time to write the code.

- The Arduino IDE has a lot of completed code examples that you can choose from.

- From the Arduino IDE, open

File→ Examples→ Liquid Crystal→ Hello World

- We need to edit the code a little bit, for the first line

LiquidCrystal lcd (12, 11, 5, 4, 3, 2);

do the following:

LiquidCrystal lcd (7, 8, 9, 10, 11, 12);

- Upload this code on the Arduino board.

```
//Example_12_LCD_16x2
#include <LiquidCrystal.h>
LiquidCrystal lcd(7, 8, 9, 10, 11, 12);
void setup()
{
lcd.begin(16, 2);
lcd.print("hello, world!");
}
void loop()
{
```

lcd.setCursor(0, 1);

lcd.print(millis()/1000);}

- This is what you will get after wiring and coding:

- Also you can change the display brightness using the potentiometer.

- You may choose any type or color from different types of character LCDs on the market.

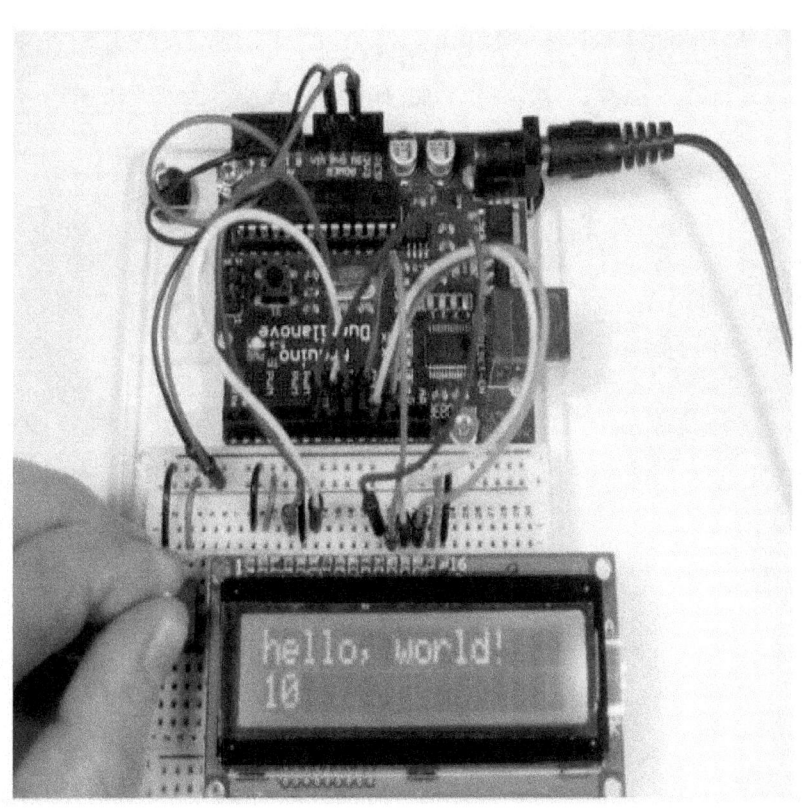

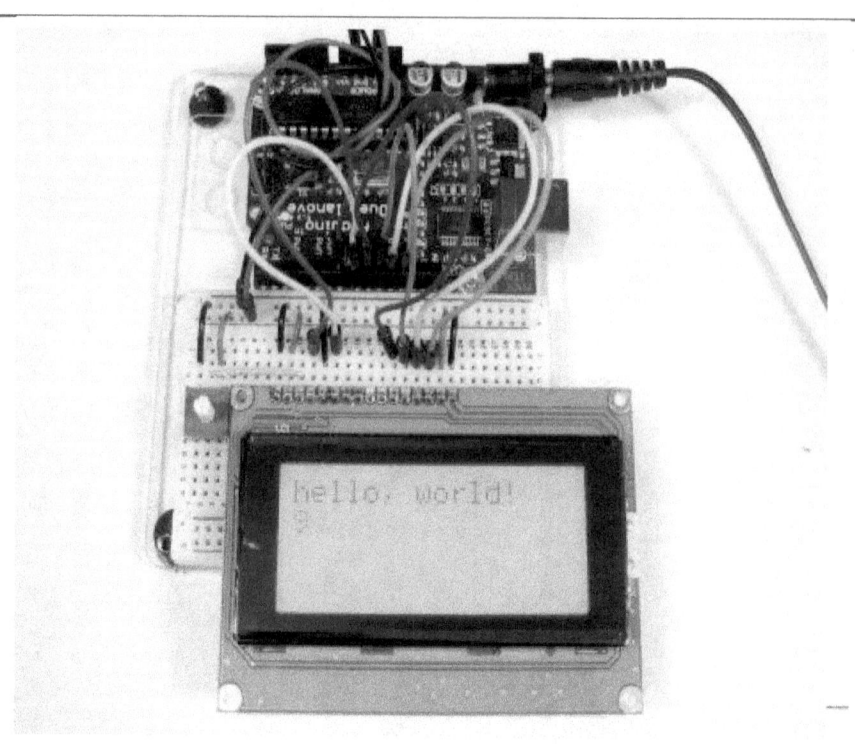

- For more examples of LCD projects visit:

https://www.arduino.cc/en/Tutorial/HelloWorld?from=Tutorial.LiquidCrystal

Connect the Keypad with the Arduino

- We consider the keypad as a very important type of advanced input device that you can find easily in a lot of applications, such as the telephone, the keyboard, the control panel on an elevator, and so on...

- There are many types of keypads. They differ in size and number of characters, and in some keypads you can find extra symbols like star (*), the Baum symbol (#), or maybe English characters such as A,D, or F

- The most popular keypad sizes are 4x4 and 4x3.

 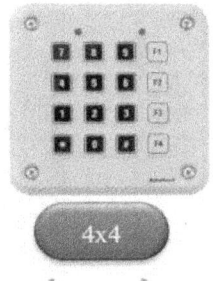 

- There are some special keypads that are more flexible and thinner that are made from certain materials, and it's very cheap as well.

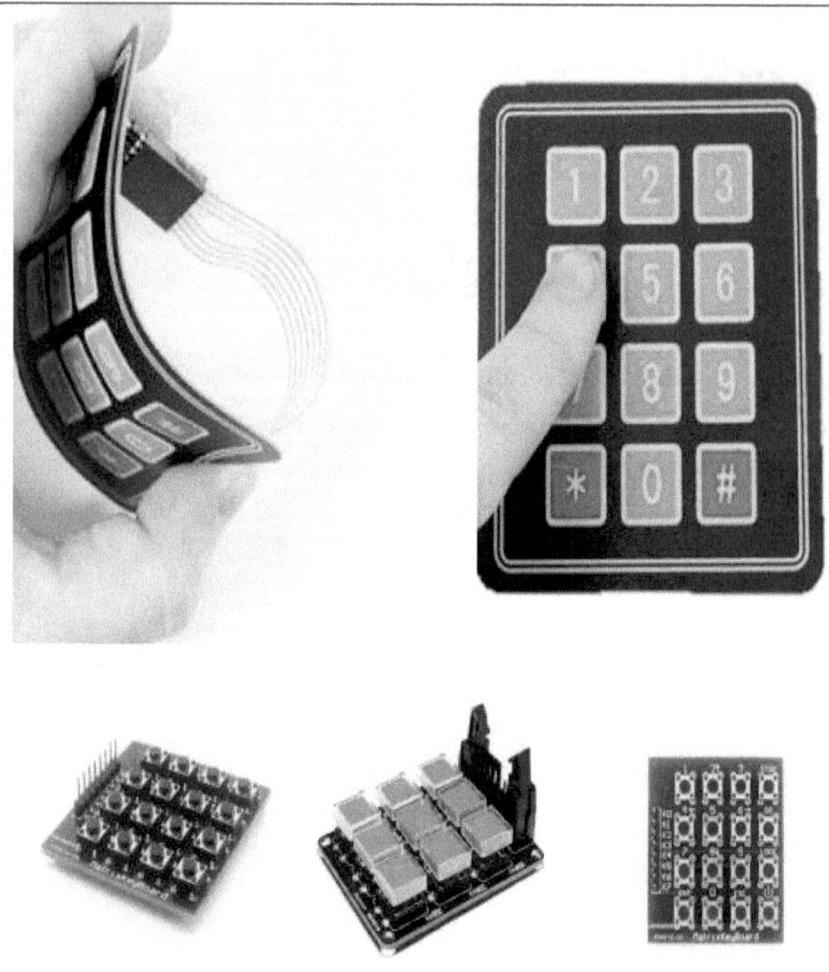

The specifications of the 3x4 keypad

- Weight: 7.5 gram

- Keypad dimensions: 70mm x 77mm x 1mm (2.75"x 3" x 0.035")

- Length of cable + connector: 85mm

- 7-pin 0.1" pitch connector

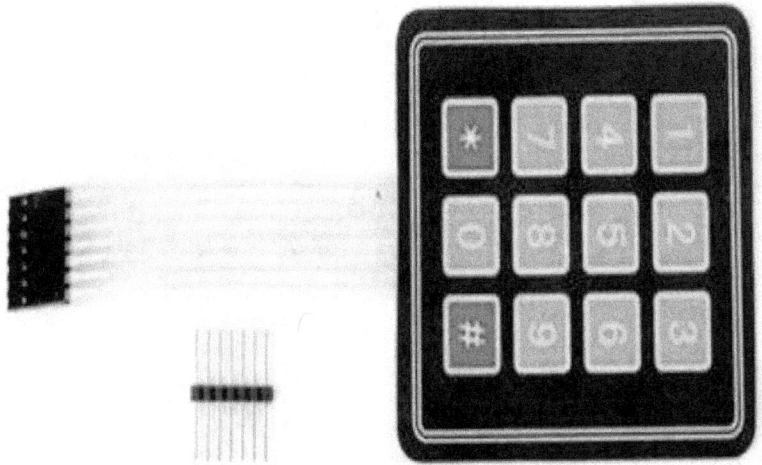

Example 11: Using the keypad with Arduino (Components)

- 3x4 keypad

- Pin headers (7)

- Arduino UNO Board

- Breadboard

Example 11: Using the keypad with Arduino (Wiring)

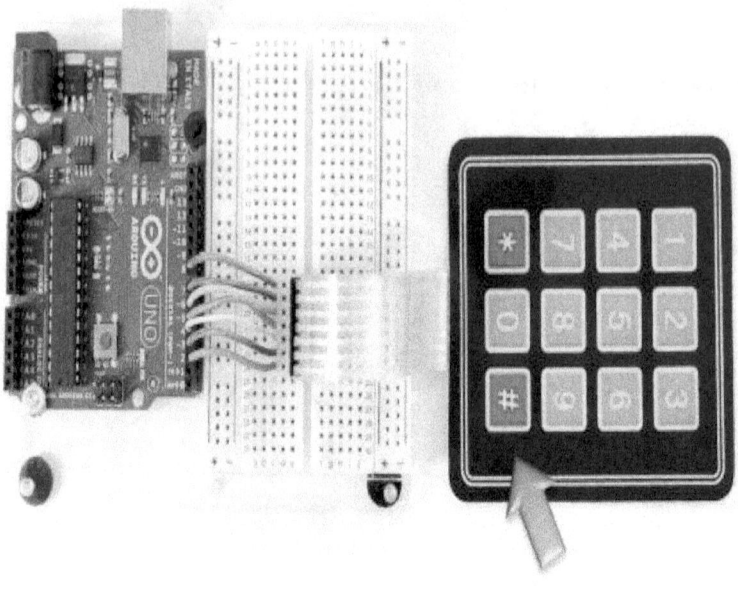

- Connect pins number 2, 3, 4, 5, 6, 7, and 8 on the Arduino to the pins on the keypad, but connect the # with pin 2 on the Arduino.

- Before you start writing the code for the Arduino, you should first download the keypad library from the Arduino website because it doesn't exist in the Arduino IDE.

Here is the link that you can download the library:

http://playground.arduino.cc/Code/Keypad

● Download and install the library as shown.

● It will be written as Keypad.zip, so the next step will be to extract the file and copy the files and paste it in the libraries folder, which is in the Arduino IDE folder on your computer as shown in the following picture:

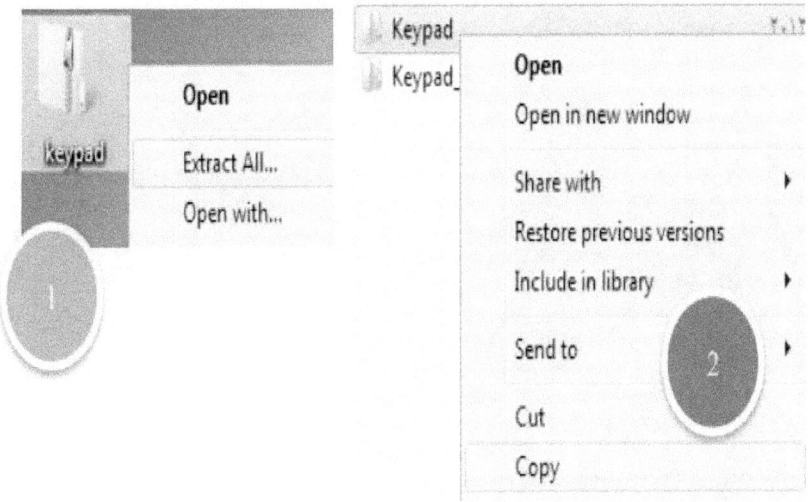

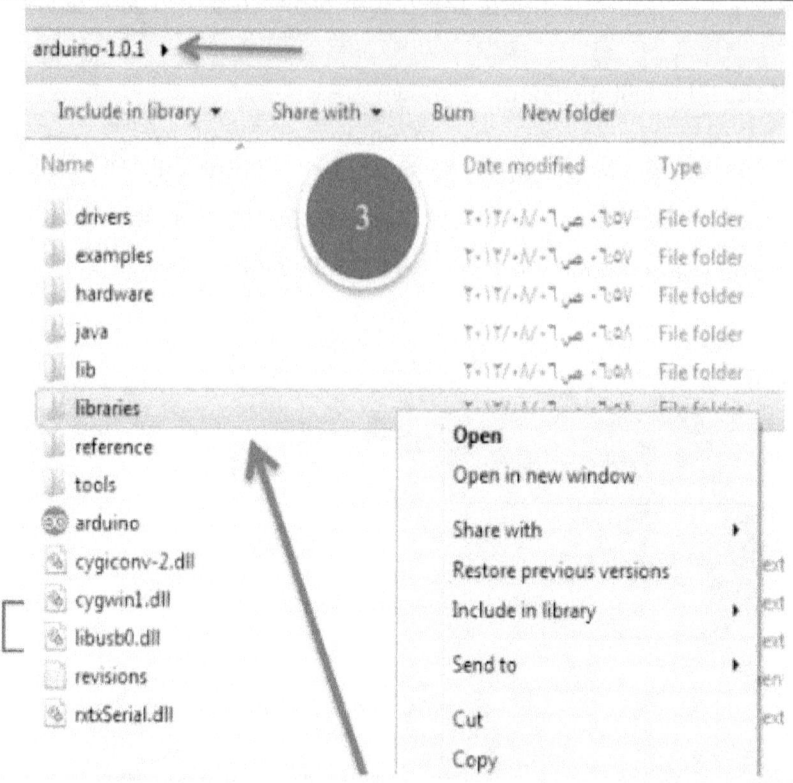

- Example11: using the keypad with Arduino (Coding)

//Example_13_Keypad_Input

#include <Keypad.h>

const byte ROWS = 4;

const byte COLS = 3;

char keys[ROWS][COLS] =

```
{
{'1','2','3'},
{'4','5','6'},
{'7','8','9'},
{'#','0','*'}
};
byte rowPins[ROWS] = {5, 4, 3, 2};
byte colPins[COLS] = {8, 7, 6};
Keypad keypad = Keypad) makeKeymap)keys(, rowPins, colPins, ROWS, COLS (;
void setup)(
{
Serial.begin)9600(;
}
void loop)(
{
char key = keypad.getKey)(;
if )key != NO_KEY( {ꞥSerial.println)key(;
```

}

}

- After writing and uploading the code on the Arduino, click on the serial monitor icon and watch will happen.

Introduction to Relays

- The relay is one of the most important components that you will see in a lot of projects, especially project with appliances.

What is a relay?

• A relay is an electromechanical device. We can imagine a relay as a switch that we can divide it into two main parts.

- **The first part is** a collided wire around the heart of the relay.

- **The second part is** an iron rectangle, and this is the switch which makes the relay in an on or off state.

The relay symbol

-
    he left side is the coil

-
    he right side is the switch

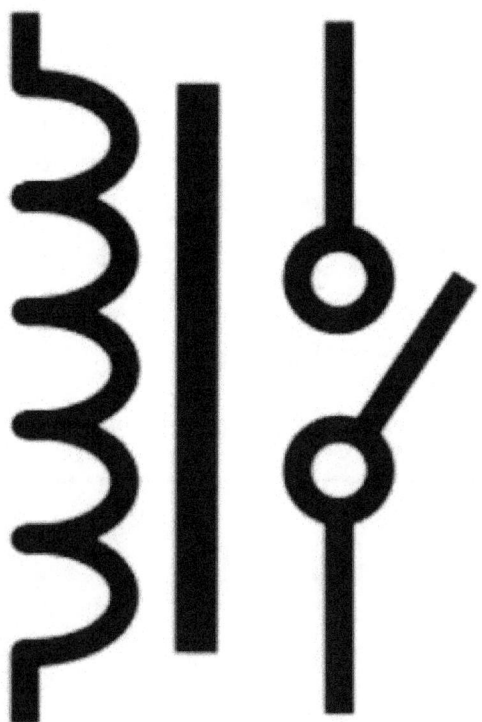

- Let's take a look at the internal design of a relay.

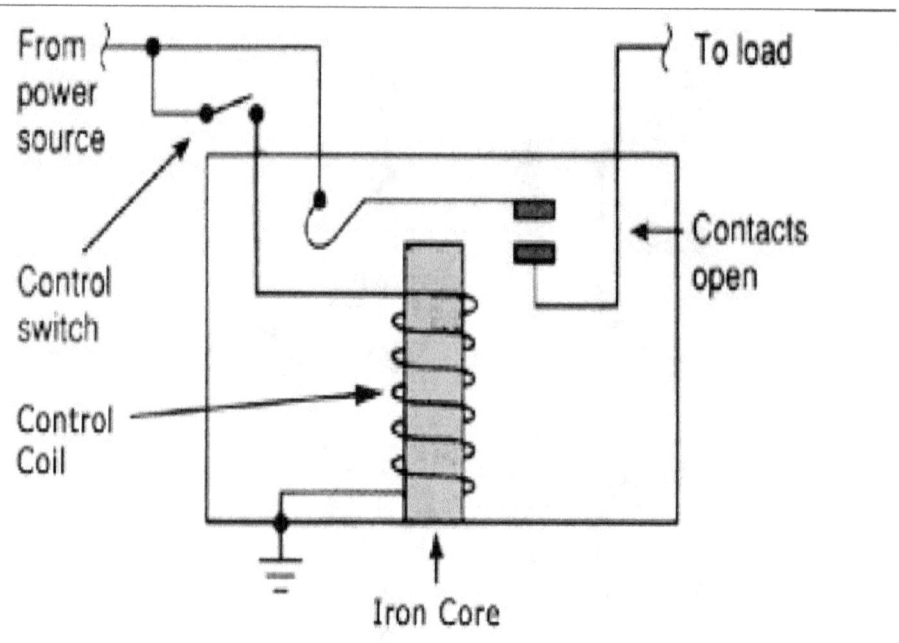

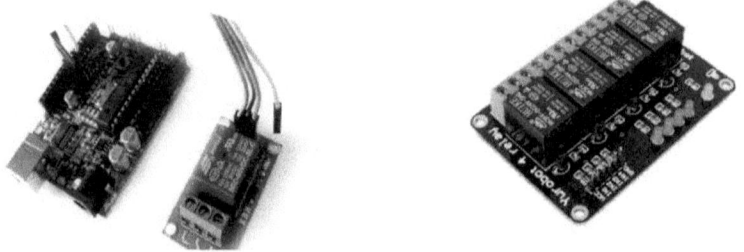

- If you want to understand more about relays go to:

https://www.sparkfun.com/tutorials/119

Questions

1.

   hat is a relay?

2. With the following parts, design an access control system:

   - Arduino board

   - Keypad 4x4

   - Proto board

   - Breadboard

   - Servo motor

   - Jumper wires

Chapter 8

Arduino shields

What you will learn in this chapter

Take a look at different types of Arduino shields

Intro to shields
- One of the most important advantage of using an Arduino is the availability of different types of shields that can integrate themselves with the Arduino.

- The idea of Arduino shields is something like your PCI cards on your PC motherboard, such as the network interface card that makes you access the Internet so easily. The Arduino shields work in the same manner.

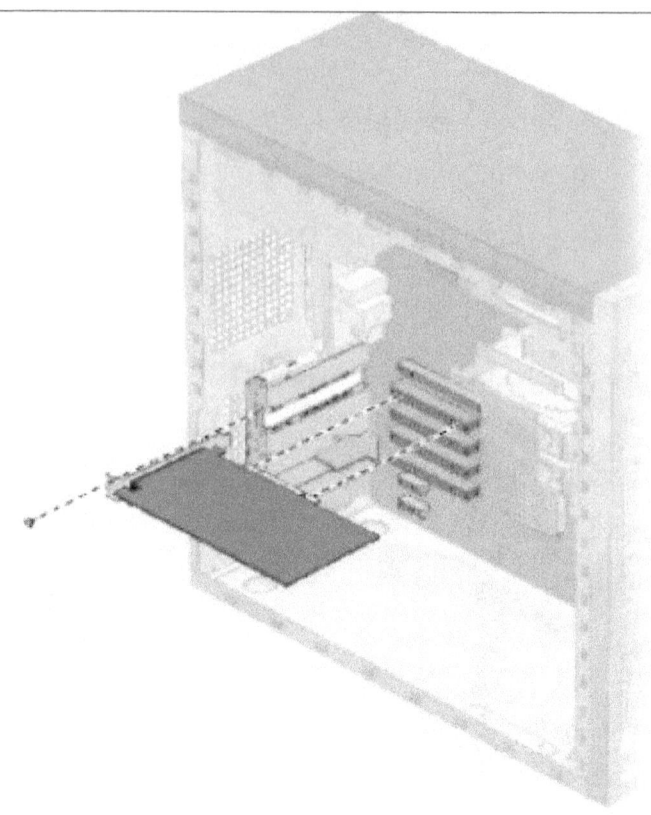

If you have any experience with microcontrollers, you can imagine the effort of connecting the microcontroller to the Internet or a local area network (in this situation you would need to build the Ethernet module from scratch), and it can take a lot of time and effort.

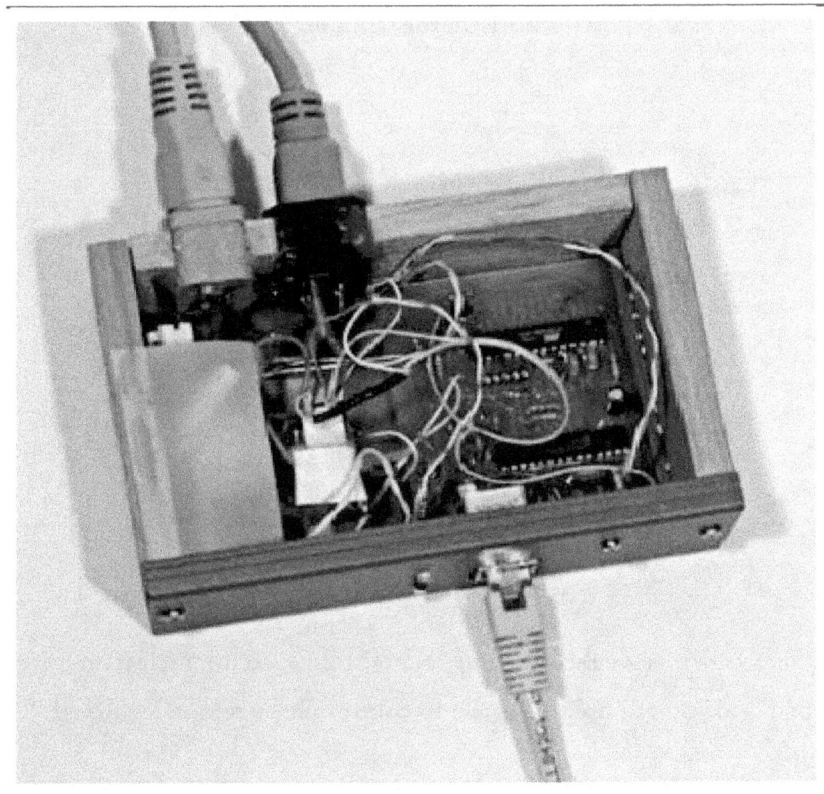

- But with Arduino, you just need to buy an Ethernet shield.

- There are many types of shields like the following examples:

Arduino Ethernet Shield

This Arduino Ethernet shield can connect the Arduino to the Internet using a CAT5 cable, and you can use this shield to control things remotely through the Internet.

- This shield has an option to add an SD card that be used to store something periodically like the temperature values.

XBee Shield

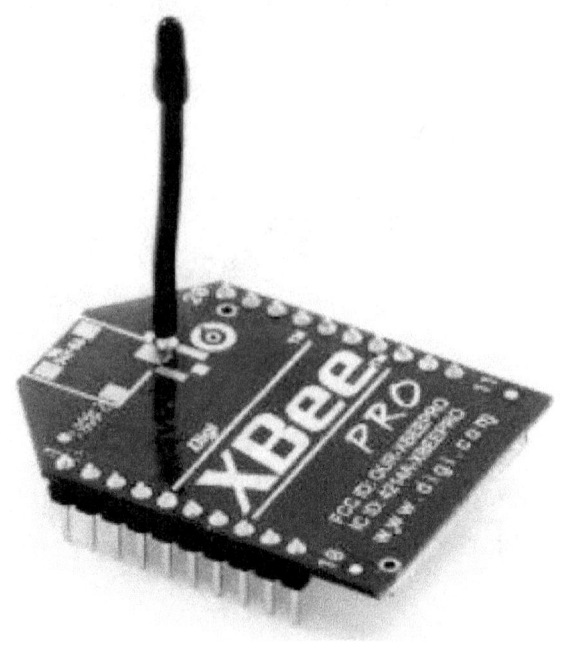

- This is the Xbee shield. It can do the same things as the Ethernet shield, but wirelessly. You can connect the Arduino to any wireless network within 100 meters.

Arduino Motor Shield

- The motor shield is used to connect different types of motors like DC motors, servo motors, and stepper motors, and you can connect three motors at the same time.

- Some versions of the motor shield allow you to connect just two motors at the same time.

- You can use this shield in projects that need motors such as robots and CNC machines.

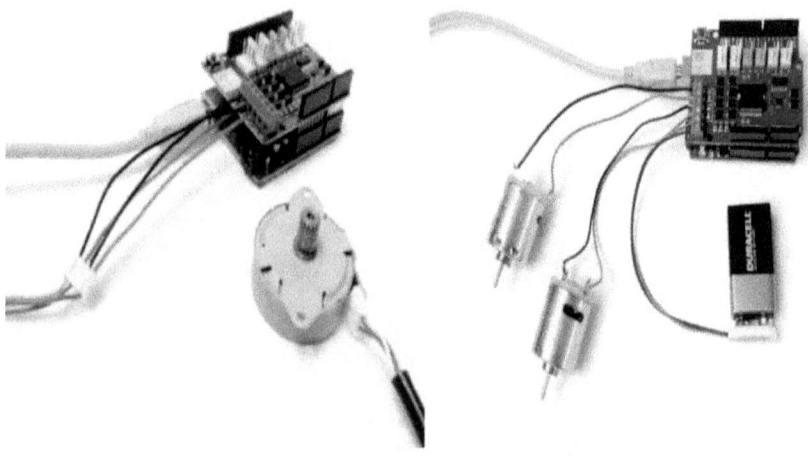

Arduino Colored Touch Screen

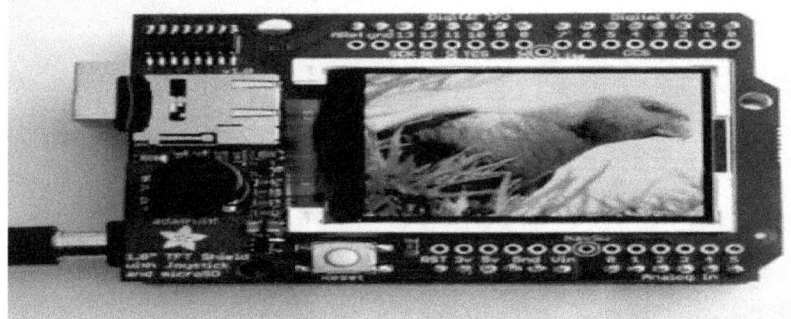

• This shield can be used in any interactive project to allow you to see some data, like photos, that the LCD cannot handle.

• There are different sizes of this kind of touch screen shield, starting from 2 inches to 4 inches.

Questions

1. What are the benefits of using Arduino shields?

2. How many motors can we use on the motor shield?

3. Describe the Xbee shield.

Chapter 9

Final Project

What you will learn in this chapter:

- Apply what you learned in one project
- Working with SD Cards

What you will need for this chapter:

- An Arduino UNO board
- USB cable
- Breadboard
- LEDs

Hardware and software requirements

In this part you will need the following components:

- An Arduino UNO board

- Arduino Ethernet shield

- USB cable

Before you start assembling the Arduino and the Ethernet shield, write down the MAC address of the shield, which is written on the back of the shield.

Hardware configuration

- The hardware configuration for this part is very simple. At this point, you should already have the Ethernet shield which is connected to the Arduino UNO board.

- Now plug the Ethernet cable into the Ethernet shield and the USB cable to the Arduino board and your computer.

- Connect the Ethernet cable to the main router of your home. Usually, you will have a Wi-Fi router in your home, which you use to enable Wi-Fi connectivity for your computer or other devices in your home network. This router should also have ports that you could connect with your Ethernet cable.

- The advantages of using this method is that your Ethernet shield will automatically get an IP address, so you can easily access the Internet in case your router is configured for DHCP which is depend on your router you use

- In case you don't have a router, you can just connect the Ethernet cable to your computer, but sharing the Internet with your Arduino board will be very complex.

Test the connection

Start the Arduino sketch with the following:

Download the Ethernet library code from Arduino.com.

#include <SPI.h>

#include <Ethernet.h>

Byte mac [] = {0x80, 8XA2, 0xDA, 0x0E, 0xFE, 0x40} //write you mac address

/*you must define the mac address to test the connection

We will test the connect by grabbing a request from any simple web page.

*/

/* The web address is stored in a char variable, you can check the data types in Chapter 3*/

char server[] = "www.example.com"; // write any website

/*the Ethernet shield will get the IP of this website*/

/* let's create an instance of the Ethernet client*/

EthernetClient client;

// write the following in your setup function

void setup()
{
　If (Ehternet.begin(mac) ==0)
　{
　　Serial.println("Failed to configure the Ethernet");

```
Ethernet.begin(mac, ip);

}

Serial.begin(115200);

Serial.print("IP address: ");

Serial.println(Ethernet.localIP());

}
```

//in your loop function, we will connect to the server by calling the connect function

```
Loop()

{

    client.printlin("GET /java/host/test.html  HTTP/1.1");

    client.println("Host: www.example.com");

client.println("connection: close");

client.println();
```

// after sending the request, we will read the data from the server to check that everything is going correctly

```
while(client.connect()) {

  while(client.availabel()) {

char c = client.read();
```

Serial.print(c);

}

/*if the client is not connected, we will print the information on the serial monitor*/

If(!client.connected())

{

Serial.println();

Serial.println("disconnecting");

Client.stop();

}

}

}

After the explanation of the same code, write the code as the following:

// Include these libraries, you can download it

#include <SPI.h>

#include <Ethernet.h>

// Enter the MAC address

byte mac[] = = {0x80, 8XA2, 0xDA, 0x0E, 0xFE, 0x40};

// Define the server

char server[] = "www.example.com";

// Set an IP address

IPAddress ip(192,168,1,50);

// create an instance

EthernetClient client;

void setup() {

  // start the serial communications

  Serial.begin(115200);

  // Start the connection

```
if (Ethernet.begin(mac) == 0) {

    Serial.println("Failed to configure the Ethernet ");

    Ethernet.begin(mac, ip);

}



// Display the IP address

Serial.print("IP address: ");

Serial.println(Ethernet.localIP());



// Give the Ethernet shield a second to initialize

delay(1000);

Serial.println("Connecting...");



}


void loop()

{

    // Connect to servers
```

```
if (client.connect(server, 80)) {

  if (client.connected()) {

    Serial.println("connected");


    // Make a HTTP request:

    client.println("GET /java/host/test.html HTTP/1.1");

    client.println("Host: www.example.com");

    client.println("Connection: close");

    client.println();

  }
  else {

    // If the connect was failed

    Serial.println("connection failed");

  }


  // answer reading

  while (client.connected()) {

    while (client.available()) {
```

```
    char c = client.read();

    Serial.print(c);

  }

}


  // If the server's disconnected, stop the client:

  if (!client.connected()) {

    Serial.println();

    Serial.println("disconnecting.");

    client.stop();

  }


}


  // Repeat every 3 seconds

  delay(3000);

}
```

Now it's the time to send the data to the web server.

In the previous part we made sure that the shield is working well and connected to your network.

In this part we will do the following:

• First, we will use the temperature and humidity sensor, and install the software components to plot the data on your computer.

• Second, we will build the code that calculates the measurements and send these measurements to the web server running on your computer.

• Third, we will build the server side code.

• Finally, we will connect the database with the plotting library, so the measurements can be seen as they get out from the Ethernet shield and are stored in our database.

Let's work with the hardware. We will need the following components for this part of our project:

- DHT11 temperature and humidity sensor / LM35 Temp sensor

- 4.7 ohm resistor

- Breadboard

- Jumper wires

- Arduino UNO board

Connect the components as shown:

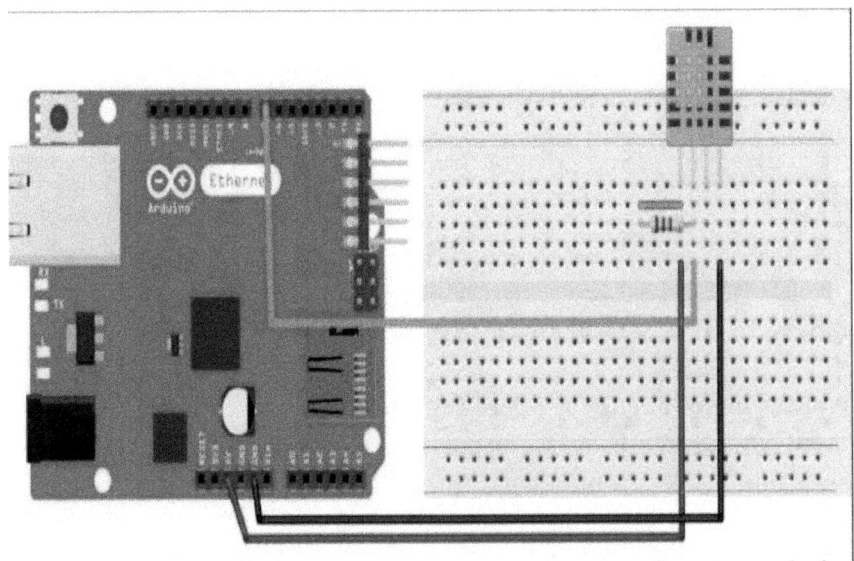

For the software components you will need the following:

- Download and include the DHT11 Library from this link:

https://playground.arduino.cc/Main/DHT11Lib

(After downloading, unzip the files and include them in the libraries folder)

- Download the plotting Library, from this link:

http://www.flotcharts.org/

- Download the database management system, we will use SQLite DMBS

http://www.sqlite.org/

- For the web server, we will use Apache.

If you are using of one these operating systems:

Windows: http://www.wampserver.com/en/

Linux: https://help.ubuntu.com/community/ApacheMySQLPHP

Mac: https://www.mamp.info/en/

Steps

• Plug the DHT11 sensor to the breadboard. Then connect pin number 1 and pin number 2 to the sensor using the 4.7k ohm resistor.

• For the power supply. Connect pin number 1 of the sensor to Arduino 5v, and pin number 4 to Arduino GND. Now you can connect the pin number 2 of the DHT sensor to Arduino pin number 7.

It is now the time to send the data to the server.

Let's build our first application using the Arduino Ethernet shield. But first, we need the IP address of your computer inside the Arduino sketch. We will also determine where the Arduino Ethernet shield should send the data.

You can find the IP address of your computer under Network Settings in the control panel on Windows.

On Linux/Mac, just start the command line and type the following command: ipconifg and click enter.

Now you are ready to build the Arduino sketch. First we will include the required libraries:

```
#include <SPI.h>
#include <Ethernet.h>
#include "DHT.h"
```

Also you need to add the Mac address, and you can find the Mac address on the back of your Arduino Ethernet shield.

```
byte mac[] = { 0x80, 0xA2, 0xDA, 0x0E, 0xFE, 0x30 }; // write you Mac address here
```

// now define the DHT11 pin on the Arduino as well as the type of the sensor

```
#define DHTPIN 7

#define DHTTYPE DHT11
```

// you can use the random () function to random data measurements if you don't have a connected DHT11 sensor

//now let's define the IP address

```
IPAddress server (192, 168, 1, 10);
```

// create an instance of the Ethernet client

```
EthernetClient client;
```

// and an instance of the DHT library

```
DHT dht(DHTPIN, DHTTYPE);
```

/*Now in the setup() function of you sketch , we will use the DHCP to get an IP */

```
Serial.begin(115200)

If (Ethernet.begin(mac) == 0)

{

  Serial.println("Failed to configure the Ethernet shield");

Ethernet.begin(ip, mac);

}
```

// write the following code to print the IP address on the serial monitor

```
  Serial.print("IP Address: ");

Serial.println(Ethernet.localIP());
```

/*in your Loop () function of the sketch, this code to take the measurements from the DHT11 sensor*/

```
float h = dht.readHumidity();

float t = dht.readTemperature();
```

/* now use this code to convert these measurements to strings*/

```
String temp = String((int) t);  // this process called casting
```

String hum = String(int) h); // if you want to learn more just google it

/* for debugging purposes, we will write the following code to print these values on the serial port. We will also check if these values are correct or not */

Serial.print("Temperature:" +temp);

Serial.pirnt("Humidity:" + hum);

/* the next thing we will do is send the data to the server. Don't panic if you cannot understand the following code, I will explain it later*/

If (client.connect(server, 80))

{

  If (client.connected()) {

Serial.println("connected");}

/* if this ran successfully, we can make the request now as in the previous part. But in this part we will use the GET request. You can search for the difference between the GET and the POST request, so now enter the IP address of your computer by using this code   */

Client.println("GET/datalogger/datalogger.php?temp=" + temp)+ "&hum=" +hum +"HTTP/1.1");

client.println("Host: 192.168.1.100");

client.println("Connection: close");

client.println();

// then write the next code to read data from the server

while (client.available()) {

while(client.available()){

char c = client.read();

Serial.print(c) ;

}

}// now you can close the connection if the client is not connected to the server

If (!client.connected()) {

Serial.println();

Serial.println("disconnecting");

Client.stop(); delay(1000); // 1 second}

- **This is the whole code for this part:**

// Include libraries

#include <SPI.h>

#include <Ethernet.h>

#include "DHT.h"

// Enter a MAC address for your controller below.

byte mac[] = { 0x80, 0xA2, 0xDA, 0x0E, 0xFE, 0x40 };

// DHT11 sensor pins

#define DHTPIN 7

#define DHTTYPE DHT11

// Set the static IP address for your Arduino board

IPAddress ip(192,168,1,60);

// IP address of your computer

IPAddress server(192,168,1,100);

// Initialize the Ethernet client, an instance of the Ethernet client

EthernetClient client;

```
// DHT instance
DHT dht(DHTPIN, DHTTYPE);

void setup() {

  // Open serial communications
  Serial.begin(115200);

  // Start the Ethernet connection
  if (Ethernet.begin(mac) == 0) {
    Serial.println("Failed to configure Ethernet using DHCP");
    Ethernet.begin(mac, ip);
  }

  // Display IP
  Serial.print("IP address: ");
  Serial.println(Ethernet.localIP());
```

```
  // Give the Ethernet shield a second to initialize

  delay(1000);

  Serial.println("Connecting...");



}


void loop()

{


  // Measure the humidity & temperature

  float h = dht.readHumidity();

  float t = dht.readTemperature();


  // Transform to String

  String temp = String((int) t);

  String hum = String((int) h);


  // Print on Serial monitor
```

```
Serial.println("Temperature: " + temp);

Serial.println("Humidity: " + hum);


// Connect to server

if (client.connect(server, 80)) {

  if (client.connected()) {

    Serial.println("connected");


    // Make a HTTP request:

    client.println("GET /datalogger/datalogger.php?temp=" + temp + "&hum=" + hum + " HTTP/1.1");

    client.println("Host: 192.168.1.100");

    client.println("Connection: close");

    client.println();

  }
  else {
    // If you didn't get a connection to the server
    Serial.println("connection failed");
```

```
  }



  // Read the answer
  while (client.connected()) {
    while (client.available()) {
      char c = client.read();
      Serial.print(c);
    }
  }



  // If the server's disconnected, stop the client:
  if (!client.connected()) {
    Serial.println();
    Serial.println("disconnecting.");
    client.stop();
  }



}
```

// Repeat every second

delay(1000);

}

In this part we will log the data in the data base.

We are now going to use PHP to build the server for our project. If you are beginner in PHP, you can check the following resource to learn the basics:

http://php.net/manual/en/tutorial.phpw

First, we will see the content of the datalogger.php file. This file will deal with the incoming requests from the Arduino board, then log the data in the database, and answer using a simple message. Note that this file has to be in a folder with the name (datalogger) on the web server.

- Let's write the code:

$temperature = intval($_GET{"temp"});

$humidity = intval($_GET["hum"]);

- We will initiate the connection with the database:

$db = new SQLite3('database.db');

If you are not familiar with the SQL commands, just go to this website:

https://www.w3schools.com/SQL/deFault.asp

• Now let's create the database columns: a unique ID that will be generated by SQLite, the timestamp column to know when the measurement was made, and the temperature and humidity data. This is done by using the following code:

$db->exec('CREATE TABLE IF NOT EXISTS measurements (id INTEGER PRIMARY KEY,
timestamp TIMESTAMP DEFAULT CURRENT_TIMESTAMP NOT NULL, temperature
INTEGER, humidity INTEGER);');

// if you are using more sensor like the light sensor, you will need to add more fields

/*now we can insert the data as a new row in the database. Since SQLite will add the ID and timestamp, we will add the temperature and the humidity */

$db->exec("INSERT INTO measurements (temperature, humidity) VALUES

('$temperature', '$humidity');");

/* to check that the data was recorded correctly, you can simply create a file readout.php* which will read the data from the database /

$db = new SQLite3(' database.db');

/*and now we will write the query to the database to get the data we want */

$results = $db->query('SELECT id, timestamp, temperature, humidity FROM measurements');

/* now we will use PHP to parse this variable that contains all the result */

while($row = $results->fetchArray())

{$dataset[] = array(strtotime($row['timestamp']) * 1000,$row['temperature']);}

// the final step is to print out the formatted data in the JSON format

Echo json_encode($dataset);

/* If you want to learn more about JSON

Visit http://json.org/

*/

/* Displaying the results */

/* we are now going to use the data in our database and display it on a graph. For this task we will use a JavaScript library, called flot, which is already included in our code. This library provides nice functions to plot the data on web pages, also it can plot the data in real time */

/* everything will happen inside an HTML file called plot.html. We will see the most important pieces of the code here*/

```
<script src="flot/jquery.js"></script>

<script src="flot/jquery.flot.js"></script>

<script src="flot/jquery.flot.time.js"></script>

<div id="placeholder" style="width:800px; height:450px;"></div>
```

/* if you want to learn more about the JavaScript

You can check out this link

https://javascript.info/

*/

- Because we have the timestamps as the x-axis, we need to determine that the data for this position is a specific time, and that we want to display this format in hours, minutes, and seconds:

```
var options = {

xaxis: {

mode: "time",

timeformat: "%H:%M:%S"

}
```

};

• We also need to receive the data every time we call the script. This will be done by an AJAX call to the PHP file.

$.ajax({

url: "readout.php",

type: "GET",

dataType: "json",

success: onDataReceived

})

// this code used in JavaScript to define function like the loop function in Arduino

   function update() {

function onDataReceived(series) {

var data = [];

data.push(series);

$.plot("#placeholder", data, options);

}

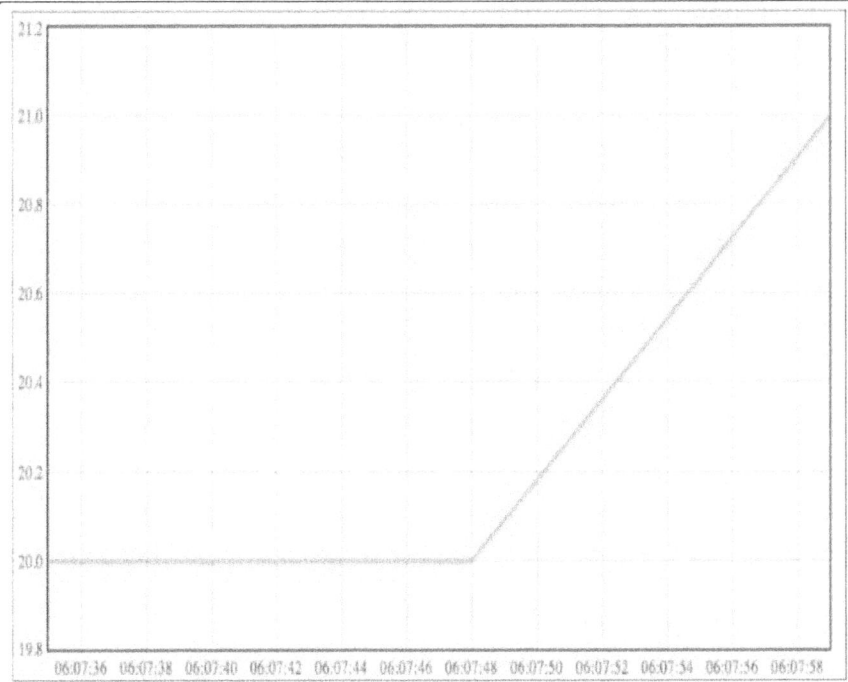

- This is what you will see:

The whole code for this part:

// web client code

// Include libraries

#include <SPI.h>

```
#include <Ethernet.h>
#include "DHT.h"

// Enter a MAC address for your controller below.
byte mac[] = { 0x90, 0xA2, 0xDA, 0x0E, 0xFE, 0x40 };

// DHT11 sensor pins
#define DHTPIN 7
#define DHTTYPE DHT11

// Set the static IP address for your board
IPAddress ip(192,168,1,50);

// IP address of your computer
IPAddress server(192,168,1,100);

// Initialize the Ethernet client
EthernetClient client;
```

```
// DHT instance
DHT dht(DHTPIN, DHTTYPE);

void setup() {

  // Open serial communications
  Serial.begin(115200);

  // Start the Ethernet connection
  if (Ethernet.begin(mac) == 0) {
    Serial.println("Failed to configure Ethernet using DHCP");
    Ethernet.begin(mac, ip);
  }

  // Display IP
  Serial.print("IP address: ");
  Serial.println(Ethernet.localIP());
```

```
// Give the Ethernet shield a second to initialize

delay(1000);

Serial.println("Connecting...");

}


void loop()

{


// Measure the humidity & temperature

float h = dht.readHumidity();

float t = dht.readTemperature();


// Transform to String

String temp = String((int) t);

String hum = String((int) h);
```

```
// Print on Serial monitor

Serial.println("Temperature: " + temp);

Serial.println("Humidity: " + hum);


// Connect to server

if (client.connect(server, 80)) {

  if (client.connected()) {

    Serial.println("connected");


    // Make a HTTP request:

    client.println("GET /datalogger/datalogger.php?temp=" + temp + "&hum=" + hum + " HTTP/1.1");

    client.println("Host: 192.168.1.100");

    client.println("Connection: close");

    client.println();

  }
  else {

    // If you didn't get a connection to the server
```

```
    Serial.println("connection failed");
}


// Read the answer
while (client.connected()) {
  while (client.available()) {
    char c = client.read();
    Serial.print(c);
  }
}


// If the server's disconnected, stop the client:
if (!client.connected()) {
  Serial.println();
  Serial.println("disconnecting.");
  client.stop();
}
```

```
  }


  // Repeat every second
  delay(1000);
}
// this is the PHP code
error_reporting(E_ALL);
ini_set("display_errors", 1);
// Check that data is present
if (isset($_GET["temp"]) && isset($_GET["hum"])) {
    // Get data
    $temperature = intval($_GET["temp"]);
    $humidity = intval($_GET["hum"]);
    // Create DB instance
    $db = new SQLite3('database.db');
    // Create new table if needed
    $db->exec('CREATE TABLE IF NOT EXISTS measurements (id INTEGER PRIMARY KEY, timestamp TIMESTAMP DEFAULT
```

CURRENT_TIMESTAMP NOT NULL, temperature INTEGER, humidity INTEGER);');

```
        // Store data in DB

        if($db->exec("INSERT INTO measurements (temperature, humidity) VALUES ('$temperature', '$humidity');")){

            echo "Data received";

        }

        else { echo "Failed to log data";            }}?>
```

//this is the plot code

```html
<!doctype html>

<html lang="en">

<head>

 <meta charset="utf-8">

 <title>Temperature readout</title>

 <script language="javascript" type="text/javascript" src="flot/jquery.js"></script>

 <script language="javascript" type="text/javascript" src="flot/jquery.flot.js"></script>

 <script language="javascript" type="text/javascript" src="flot/jquery.flot.time.js"></script>
```

```
</head>

<body>


<div id="placeholder" style="width:800px; height:450px;"></div>


<script>

$(function () {

        var options = {

                xaxis: {

                        mode: "time",

                        timeformat: "%H:%M:%S"

                }

        };

        // Update plot

        function update() {

                // Get data

                $.ajax({
```

```
            url: "readout.php",
            type: "GET",
            dataType: "json",
            success: onDataReceived
        });

        // Plot data
        function onDataReceived(series) {
            var data = [];
            data.push(series);
            $.plot("#placeholder", data, options);
        }

        // Time interval between updates
        setTimeout(update, 10);
    }

    // Update the plot
    update();
});
</script>
```

</body>

</html>

//this is the read out script code

<?php

// Show errors

error_reporting(E_ALL);

ini_set("display_errors", 1);

```php
// Open database
$db = new SQLite3('database.db');

// Set default timezone
date_default_timezone_set('America/Los_Angeles');

// Get data
$results = $db->query('SELECT id, timestamp, temperature, humidity FROM measurements');

// Parse data
while($row = $results->fetchArray())
{
    $dataset[] = array(strtotime($row['timestamp']) * 1000,$row['temperature']);
}

// Return data
echo json_encode($dataset);?>

// the data logger code
?php

error_reporting(E_ALL);

ini_set("display_errors", 1);
```

```php
// Check that data is present

if (isset($_GET["temp"]) && isset($_GET["hum"])) {

    // Get data

    $temperature = intval($_GET["temp"]);

    $humidity = intval($_GET["hum"]);

    // Create DB instance

    $db = new SQLite3('database.db');

    // Create new table if needed

    $db->exec('CREATE TABLE IF NOT EXISTS measurements (id INTEGER PRIMARY KEY, timestamp TIMESTAMP DEFAULT CURRENT_TIMESTAMP NOT NULL, temperature INTEGER, humidity INTEGER);');

    // Store data in DB
```

```
    if($db->exec("INSERT INTO measurements (temperature,
humidity) VALUES ('$temperature', '$humidity');")) {

        echo "Data received";

    }

    else {

        echo "Failed to log data";

    }

}
?>
```

In this part we are going to deal with the SD cards to store the data, it should also be in FAT32 format.

These are the parts you will need for our final part:

- The Arduino UNO

- The Arduino Ethernet shield

- The DHT11 sensor

- A MicroSD card

- A breadboard

- Jumper wires

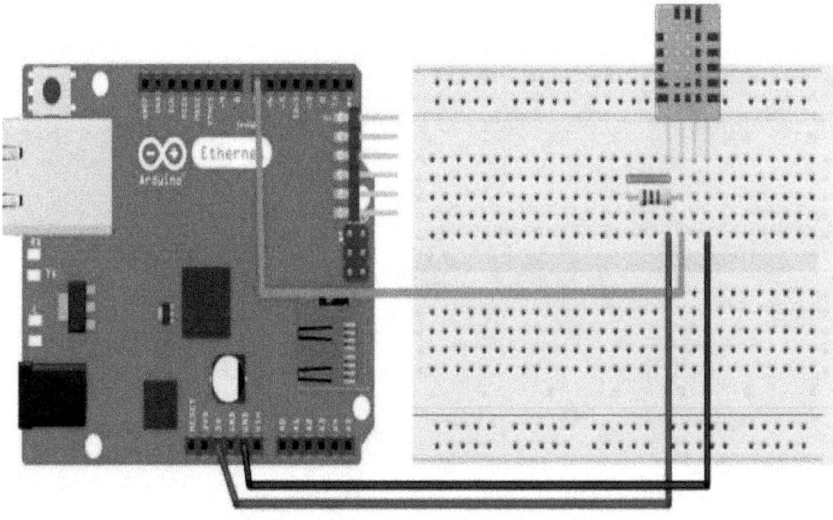

- Now connect the Arduino and the SD card as shown.

- This is the whole code for this part:

//datalogger code

// Include libraries

#include "DHT.h"

#include <SD.h>

```
#include <Time.h>

#include <Ethernet.h>

#include <EthernetUdp.h>

#include <SPI.h>


// DHT11 sensor pins

#define DHTPIN 7

#define DHTTYPE DHT11


// Enter a MAC address for your controller below.

byte mac[] = { 0x90, 0xA2, 0xDA, 0x0E, 0xFE, 0x40 };


// Chip select pin

const int chipSelect = 4;


// NTP Server

IPAddress timeServer(132, 163, 4, 101);

const int timeZone = 1;
```

```
// Create UDP server

EthernetUDP Udp;

unsigned int localPort = 8888;


// DHT instance

DHT dht(DHTPIN, DHTTYPE);


void setup() {

  // Open serial communications

  Serial.begin(9600);


  // Start Ethernet

  if (Ethernet.begin(mac) == 0) {

    // no point in carrying on, so do nothing forevermore:

    while (1) {

      Serial.println("Failed to configure Ethernet using DHCP");
```

```
    delay(10000);

  }
}


// Init SD card

Serial.print("Initializing SD card...");

pinMode(10, OUTPUT);


if (!SD.begin(chipSelect)) {

  Serial.println("Card failed, or not present");

  // don't do anything more:

  return;

}

Serial.println("card initialized.");


// Initialize DHT sensor

dht.begin();
```

```
// Print server info
Serial.print("IP number assigned by DHCP is ");
Serial.println(Ethernet.localIP());
Udp.begin(localPort);
Serial.println("waiting for sync");
setSyncProvider(getNtpTime);
}


void loop() {

  // Measure the humidity & temperature
  float h = dht.readHumidity();
  float t = dht.readTemperature();

  // Transform to String
  String temp = String((int) t);
  String hum = String((int) h);
```

```
// Format time
String log_time = String(day()) + "/" +
String(month()) + "/" + String(year()) + " " +
String(hour()) + ":" + String(minute()) + ":" +
String(second());


// Make a string for assembling the data to log
String dataString = log_time + "," + temp + "," + hum;


// Open file
File dataFile = SD.open("datalog.txt", FILE_WRITE);


// Write data to file
if (dataFile) {
  dataFile.println(dataString);
  dataFile.close();
  Serial.println(dataString);
```

```
    }
  else {
    Serial.println("error opening datalog.txt");
  }


  // Repeat every 10 seconds
  delay(10000);
}



const int NTP_PACKET_SIZE = 48; // NTP time is in the first 48 bytes of message

byte packetBuffer[NTP_PACKET_SIZE]; //buffer to hold incoming & outgoing packets


time_t getNtpTime()
{
  while (Udp.parsePacket() > 0) ; // discard any previously received packets
  Serial.println("Transmit NTP Request");
  sendNTPpacket(timeServer);
```

```
uint32_t beginWait = millis();

while (millis() - beginWait < 1500) {

  int size = Udp.parsePacket();

  if (size >= NTP_PACKET_SIZE) {

    Serial.println("Receive NTP Response");

    Udp.read(packetBuffer, NTP_PACKET_SIZE);  // read packet into the buffer

    unsigned long secsSince1900;

    // convert four bytes starting at location 40 to a long integer

    secsSince1900 =  (unsigned long)packetBuffer[40] << 24;

    secsSince1900 |= (unsigned long)packetBuffer[41] << 16;

    secsSince1900 |= (unsigned long)packetBuffer[42] << 8;

    secsSince1900 |= (unsigned long)packetBuffer[43];

    return secsSince1900 - 2208988800UL + timeZone * SECS_PER_HOUR;

  }

}

Serial.println("No NTP Response :-(");

return 0; // return 0 if unable to get the time
```

}

// send an NTP request to the time server at the given address

void sendNTPpacket(IPAddress &address)

{

// set all bytes in the buffer to 0

memset(packetBuffer, 0, NTP_PACKET_SIZE);

// Initialize values needed to form NTP request

// (see URL above for details on the packets)

packetBuffer[0] = 0b11100011;   // LI, Version, Mode

packetBuffer[1] = 0;   // Stratum, or type of clock

packetBuffer[2] = 6;   // Polling Interval

packetBuffer[3] = 0xEC; // Peer Clock Precision

// 8 bytes of zero for Root Delay & Root Dispersion

packetBuffer[12] = 49;

packetBuffer[13] = 0x4E;

packetBuffer[14] = 49;

packetBuffer[15] = 52;

```
    // all NTP fields have been given values, now
    // you can send a packet requesting a timestamp:
    Udp.beginPacket(address, 123); //NTP requests are to port 123
    Udp.write(packetBuffer, NTP_PACKET_SIZE);
    Udp.endPacket();
}
```

```php
// the read out.php code
<?php
error_reporting(E_ALL);
ini_set("display_errors", 1);
$db = new SQLite3('database.db');
$results = $db->query('SELECT id, timestamp, temperature, humidity FROM measurements');
while($row = $results->fetchArray())
{
    $dataset[] = array(strtotime($row['timestamp']) * 1000,$row['temperature']);
```

}

echo json_encode($dataset);

?>

// the plot.html code

<!doctype html>

<html lang="en">

<head>

  <meta charset="utf-8">

  <title>Temperature readout</title>

  <script language="javascript" type="text/javascript" src="flot/jquery.js"></script>

  <script language="javascript" type="text/javascript" src="flot/jquery.flot.js"></script>

  <script language="javascript" type="text/javascript" src="flot/jquery.flot.time.js"></script>

</head>

<body>

```
<div id="placeholder" style="width:800px; height:450px;"></div>


<script>

$(function () {

        var options = {

                xaxis: {

                        mode: "time",

                        timeformat: "%H:%M:%S"

                }

        };

        // Update plot

        function update() {

                // Store data

                $.ajax({

                        url: "datalogger.php",

                        type: "GET",

                });
```

```
// Get data

$.ajax({

    url: "readout.php",

    type: "GET",

    dataType: "json",

    success: onDataReceived

});

// Plot data

function onDataReceived(series) {

    var data = [];

    data.push(series);

    $.plot("#placeholder", data, options);

}

// Time interval between updates

setTimeout(update, 10);

    }

// Update the plot

update();
```

```
});

</script>



</body>

</html>
```

```
// the datalogger.php code

<?php

error_reporting(E_ALL);

ini_set("display_errors", 1);

// Arduino board

$url = 'http://192.168.1.103';

// Get cURL resource

$curl = curl_init();

// Set some options - we are passing in a useragent too here

curl_setopt_array($curl, array(

    CURLOPT_RETURNTRANSFER => 1,
```

    CURLOPT_URL => $url,

));

// Send the request & save response to $resp

$resp = curl_exec($curl);

// Close request to clear up some resources

curl_close($curl);

// Get data

$json = json_decode($resp, true);

$temperature = intval($json["temperature"]);

$humidity = intval($json["humidity"]);

// Create DB instance

$db = new SQLite3('database.db');

// Create new table if needed

$db->exec('CREATE TABLE IF NOT EXISTS measurements (id INTEGER PRIMARY KEY, timestamp TIMESTAMP DEFAULT CURRENT_TIMESTAMP NOT NULL, temperature INTEGER, humidity INTEGER);');

// Store data in DB

```php
$db->exec("INSERT INTO measurements (temperature, humidity) VALUES ('$temperature', '$humidity');");


// Answer

echo "Data received";

?>
```

www.ingramcontent.com/pod-product-compliance
Lightning Source LLC
Chambersburg PA
CBHW050207230526
45470CB00001B/271